Netflix Recommends

The publisher and the University of California Press Foundation gratefully acknowledge the generous support of the Robert and Meryl Selig Endowment Fund in Film Studies, established in memory of Robert W. Selig.

Netflix Recommends

ALGORITHMS, FILM CHOICE,
AND THE HISTORY OF TASTE

Mattias Frey

UNIVERSITY OF CALIFORNIA PRESS

University of California Press
Oakland, California

Library of Congress Cataloging-in-Publication Data

Names: Frey, Mattias, author.
Title: Netflix recommends : algorithms, film choice, and the history of taste /
    Mattias Frey.
Description: Oakland, California : University of California Press, [2021] | Includes
    bibliographical references and index.
Identifiers: LCCN 2021006515 (print) | LCCN 2021006516 (ebook) |
    ISBN 9780520382381 (cloth) | ISBN 9780520382046 (paperback) |
    ISBN 9780520382022 (epub)
Subjects: LCSH: Netflix (Firm) | Streaming video—Social aspects—United
    States. | Recommender systems (Information filtering)—Social aspects.
Classification: LCC HD9697.V544 N48236 2021 (print) | LCC HD9697.V544 (ebook)
    | DDC 384.55/54—dc23
LC record available at https://lccn.loc.gov/2021006515
LC ebook record available at https://lccn.loc.gov/2021006516

30   29   28   27   26   25   24   23   22   21
10   9   8   7   6   5   4   3   2   1

CONTENTS

ACKNOWLEDGMENTS

I gratefully acknowledge the support of the Leverhulme Trust, whose Philip Leverhulme Prize (PLP-2015-008) funded the majority of the underlying research contained in the following pages. Smaller research grants from the School of Arts and the Faculty of Humanities of the University of Kent provided the necessary resources to finish the project.

This book draws on nearly a decade of research, undertaken in various settings. My sincere thanks go to YouGov and to the librarians at the BFI Reuben Library and the British Library for their friendly assistance. I extend my gratitude to the participant interviewees for taking the time to demonstrate and explain their behaviors, preferences, and predilections. Roderik Smits was the ideal research assistant for this endeavor, always approaching his tasks with intelligence and sensitivity.

Although this project was conceived as a book from the start, I tested portions at various research seminars and conferences, including at BAFTSS, Università di Bologna, University of Bristol, Masaryk University Brno, University College Cork, Dartmouth College, University of Edinburgh, Finnish Film Institute, Universiteit Gent, Harvard University, HoMER @ NECS, King's College London, Liverpool John Moores University, Universität Mainz, MeCCSA, Oxford Brookes University, and SCMS.

I thank the organizers of these events for their invitations to speak about these issues and to all those who posed questions or provided suggestions along the way. These generous scholars include Sarah Atkinson, Daniël Biltereyst, Erica Carter, Alex Clayton, Paul Cooke, Ginny Crisp, Gerd Gemünden, Šárka Gmiterková, Rasmus Greiner, Noah Isenberg, Amanda Landa, Jon Lewis, Amanda Lotz, Giacomo Manzoli, Alexander Marlow-Mann, Paul McDonald, Maya Nedyalkova, Paulo Noto, Lydia Papadimitriou, Winfried Pauleit, Ondřej Pavlík, Laura Rascaroli, Rick Rentschler, Kirsi Rinne, Cecilia Sayad, David Sorfa, Peter Stanfield, Daniela Treveri Gennari, Yannis Tzioumakis, Chris Wahl, and Jonathan Wroot. To these individuals and the many others whom I do not name here: I am genuinely grateful.

Sincere thanks are due to all my colleagues in the Department of Film and Media Studies, School of Arts, and Centre for Film and Media Research at the University of Kent, and to the Heads of School during this period: Peter Stanfield, Martin Hammer, and Tamar Jeffers-McDonald. Research administrators, directors, and associate deans provided key assistance at crucial junctures. I thank Lynne Bennett, Helen Brooks, Gillian Lodge, Catherine Richardson, Michelle Secker, and Aylish Wood.

Working with the University of California Press has been extraordinary. Raina Polivka has been supportive, encouraging, efficient, and effective from start to finish. I gratefully appreciate her, Jon Dertien, Gary J. Hamel, Teresa Iafolla, Francisco Reinking, Madison Wetzell, and the rest of the staff for their tremendous professionalism and work behind the scenes.

The two external peer reviewers, Ramon Lobato and Denise Mann, provided exceptionally generous and insightful advice. The manuscript is much better because of their wisdom; the residual infelicities remain mine only.

Once again, my family, friends, and まり子 were the lamps unto my feet, the lights unto my path. One million thanks.

The following manuscript includes a few recycled sentences from my short essays "The Internet Suggests: Film, Recommender Systems, and Cultural Mediation," *Journal of Cinema and Media Studies* 59, no. 1 (2019): 163–69; and "The Ends of (German) Film Criticism: On Recurring Doomsday Scenarios and the New Algorithmic Culture," *New German Critique* 47, no. 3 (2020): 45–57.

Introduction

THE SEEDS OF THIS BOOK go back at least ten years. In those days, words hitherto familiar to me only from art galleries and craftsmen's wood shops suddenly abounded in all fields of culture and leisure. At cafés in the hip part of town, sommeliers were called wine "curators"; baristas selected and poured "artisanal" coffee from a far-flung nation, just for me. At the same time, emerging music services such as Spotify promised—for a small monthly fee or the relative inconvenience of occasional advertisement interruptions— to play an assortment of songs new to me but perfectly tailored to my taste, using cutting-edge technology and computer processing.

Of course, there has always been tech-speak and hipster language, semantic gentrification and outright snobbishness. But the marketing rhetoric seemed to want to take sides: either firmly partaking of a belief in the strength of numbers, algorithms, and computational systems to better provide cultural services—or, at the other end of the spectrum, gleefully luddite appeals to the human touch, irreverent celebrations of gatekeeping expertise, bijou or bespoke designs, and tribalist cultural distinction. These were business models and sales tales of personalization and community, discovery and satisfaction, expanded selection and immediate access. The wisdom of crowds, served to suit individual whims.

The forces of globalization and digitalization seemed to be expanding cultural choice. At the same time, the methods of *how* to whittle down these arrays of offerings to a manageable size—and *who* or *what* should guide that decision-making—came up for revision and debate.

The world of film and other audiovisual media—my professional domain and a significant part of my personal life—was hardly unaffected by these developments. Indeed, I noted my own behaviors changing. I had long been invested in a methodical rhythm to my week, month, and year of audiovisual consumption. The Thursday newspaper film reviews that informed my weekend cinema trips. The monthly film magazine that told me which home videos to buy or register on my DVD-by-post queue. The annual trip to a major A-list festival; the relative disinterest in television. Even with the rise of various technologies, from VHS and DVD to Blu-ray, there remained a comfortable familiarity about where I needed to go, what I needed to consult, and whom I should trust with my consumption decisions.

These once-steady, well-worn rituals were becoming superfluous, even silly, for many, including creatures of habit like myself. Too little time to read several reviews of a film? I could check out the Rotten Tomatoes or Metacritic composite scores, which conveniently popped up in the sidebar of internet searches for individual titles. No opportunity to travel to one of the major festivals this year—or unable to see even a small fraction of the hundreds of films on offer over ten days? This was no longer a real worry, because I could subscribe to MUBI for a curated selection of the previous year's best festival-circuit hidden gems. Suddenly wanting a thriller or to start bingeing on a new comedy series at home in my pajamas? Video-store trips or interminable waits for the postman were bygone and outmoded, because I could immediately consult Netflix and its personalized recommender system would select the best title for me. If I was unimpressed with these selections, I could access the *New York Times'* Watching site: it could calculate which of my four video on demand (VOD) subscriptions were carrying which well-reviewed films or series by genre or mood. Personally, anecdotally, it felt as if a fundamental change was in motion. But I recognized myself as an exception; after all, I had made a hobby into my profession. Did my lived experience and local observations correspond to a larger and perhaps even irrevocable shift in choice behaviors?

Among many newspaper critics, academics, cinephiles, and TV junkies, there was a widespread sense of empowerment in those halcyon days. It was a renewed feeling of mastery and control, of being listened and catered to, in the name of personalization: a treasure trove of moving images, *prêt à regarder*. Heavy users of media culture (and in particular the audiovisual stories we used to call cinema and television) were subscribing to Netflix, Amazon, MUBI, and other VOD services. Many harbored completist hopes of all film and television history available immediately with the click of a button.[1] Fanboys and fangirls predicted that cult tastes would go mainstream and surmised that media executives would henceforth pay them heed.[2]

Entrepreneurs envisioned tapping into vast new domains of value (i.e., making money) by linking niche items with niche markets.[3] Some even thought that these new avenues of sped-up, on-demand viewing would increase viewers' agency and activity, their participation and enjoyment, their self-reflection and democratic citizenship.[4] Euphoric talk of "convergence" and "choice," of "radical changes" and above all "revolution" abounded. Perhaps the most important "early narrative of digital change," according to one industry professional, "was of a democratising trend away from 'gatekeepers' restricting choice," the consumer-empowerment rhetoric of recommendation toward individual taste.[5] Cultural suggestions would surely become more objective, no longer at the mercy of the whims of pointy-headed elites.

Netflix's, YouTube's, and other internet streaming services' algorithmic recommender systems seemed to constitute the culmination of a certain internet fantasy: personalization. For many commentators, these systems—which suggest content likely to interest viewers based on their prior viewing histories—represented a fundamentally new way of connecting cultural objects and human beings. Computer scientists and business gurus swooned over the ability to scale the provision of cultural recommendation using big-data-based "collective intelligence" and "wisdom of crowds."[6] Feature writers for the *Atlantic*, *New Yorker*, and other middlebrow publications attested to the Netflix recommendation engine's superhuman qualities, its "alien" recognition of taste able to perceive deep structures and networks between seemingly disparate genres and actors, connections that humans and critics could not possibly intuit.[7]

It was not long, however, before scholars began to deflate such talk. The objections emerged from many different perspectives across a plethora of articles and books, but in general revolved around three persistent, overlapping concerns. The first interrogated the lack of transparency to algorithmic recommender systems, the ways that they surveil and invade, obfuscate and circumscribe human taste and culture itself. Critics questioned the opacity of the technology and the shady corporations that controlled it; they wondered why a certain film or series appeared as a suggestion. Are recommendations truly based on "taste" or simply pay-for-play promotions? The nontransparent, black-box quality of proprietary algorithms and the tracking methods that invisibly record viewing histories in order to suggest further videos for "people like you" reminded these thinkers of Foucault's panopticon and Deleuze's control society. A second major concern hinged on filter bubbles, cultural homogenization, gated communities, reputation silos, public sphericules, and social fragmentation. Algorithmic recommender systems, observers warned, hew too closely to previous selections and biases, inevitably leading users to consume certain products, thereby "hiding" others and affecting individuals' exposure to diversity. Leaving users in the dark about alternate choices, these devices limit expression and diversity, erode democratic access, narrow horizons of expectation, and inhibit empathy by erasing common experiences and "watercooler moments." Algorithms work to confirm, rather than develop or challenge, consumers' tastes, they reasoned, potentially leading to an atomized proliferation of house-bound non-communities of one. Finally, a third vocal criticism surmised that algorithmic recommendation, by virtue of its very form and technology, represents a hostile takeover of humanism, a hijacking of culture itself from the qualitative to the quantitative. Cultural recommendation—not to mention the livelihood of critics and educators since time immemorial—has traditionally been based on the presentation and evaluation of the Arnoldian "best which has been thought and said." In the face of algorithms, however, it risked being reduced to slack-jawed perma-bingers passively acquiescing to Netflix's advice that the next episode will begin in 5, 4, 3, 2, 1. These commentators entertained scenarios of a "datafication of identity" and a "mathematization of taste," indeed a wholesale redefinition of culture. SVOD (subscription video on demand) viewership—after all the fastest-growing mode of film and series consumption—risked substituting the Pauline Kaels and Manny Farbers of the world for AI machine learners, avatars for the economic imperatives of profit-seeking media conglomerates.[8] Across all three main planks the initial celebrations of grassroots democracy—declarations of the definitive end of gatekeepers and the overturning of cultural hierarchies—were met with equally certain complaints about dumbing down and humanity losing its monopoly on determining cultural value.[9]

In sum, reviewing the discourse surrounding recommender systems and related developments over the past ten, fifteen, twenty years reveals two competing, and largely mutually exclusive, narratives. One heralds an unprecedented era of democratic access and choice. The other proposes a scenario straight out of *Clockwork Orange* (1971): media shoveled into our eyes, a color-by-numbers operation masked by clever marketing illusions. The fronts in this debate could hardly be clearer or more diametrically opposed. Curiously, however, both the vociferous champions and the vehement critics share a common first-principle assumption: that VOD recommender systems are effective, powerful, widely used, and unprecedented.

In crucial ways, these hopes, dreams, anxieties, and nightmares remain at least as interesting as the forms and functionality of the technology itself. Algorithmic big-data processing and the internet have reinvigorated deep-seated desires about how society should be organized, and who should lead or control its opinion-leading communications apparatus. The wildly and alternately utopian and dystopian diagnoses can be explained and perhaps even justified by prime-mover and growing-pain timings, by the fact that, as Vincent Mosco and others detail, new media have over the course of history always attracted hyperbolic dream-or-nightmare talk.[10] But a quarter century after the widespread consumer embrace of the internet and the advent of Netflix and its recommender system, we no longer have the luxury of general forecasts. The initial Wild West phase of digital film and series distribution is over; the online consumption of audiovisual narratives is no longer a novelty enjoyed by elite early adopters. Having reached a "mature stage" of VOD development, largely dominated by an oligopoly of providers with culturally minded platforms operating at the fringes, it is high time to take stock of the promises and warnings about recommender systems and provide a more nuanced assessment.[11] In particular, this book squarely scrutinizes what is perhaps the most pressing of the myriad claims made about VOD recommendation engines: AI's supposed hostile takeover of cultural suggestion and indeed humanistic culture itself, algorithmic systems' putatively unprecedented re-mastery of taste.

In explaining digital audiovisual culture and its systems of presenting and recommending films and series, the discussion in this book revolves around VOD platforms, including and especially the world market leader, Netflix. Most who have written on this subject—and there are many, from business columnists and tech prophets to film critics and media scholars—see Netflix and competitors above all as the new digital distribution. A central conceit of this book, however, is to consider VOD services not only as distributors, exhibitors, or producers of moving images. Rather, we need to consider them as symptoms and enablers of cultural evaluation and taste, and within ensembles of recommendation forms, heuristics, and repertoires. VOD may be important for its internet streaming technologies, instant access and narrowcasting, and the content that has sprung from its creators and commissioners, from *House of Cards* (2013–2018) to *The Handmaid's Tale* (2017–present). But, I submit, it may be even more important for the routines of recommendation, presentation, gatekeeping, and ultimately taste that such technologies enable and incentivize. Yes, the question of what we are watching these days, and by what means, is an important issue. And yet the question of *how we come to watch what we watch* is equally crucial, and deserving of more serious, careful, and systematic scrutiny.

Indeed, we might begin to question to what extent these services have come to usurp basic functions of criticism and displace traditional media consumption routines. It is clear that algorithmic recommender systems have become an increasingly prevalent and thus important means to learn about, choose, and consume moving images. But do they also represent, as commentators posit, a fundamentally new form of recommendation, a system and logic that offers an irrevocably novel idea of culture and taste?

FROM THE PRODUCER TOWARD THE USER, FROM NOVELTY TOWARD MEDIA ARCHAEOLOGY

Twenty-five years ago, people saw films and series primarily in cinemas, on terrestrial and cable television, or on their VCR. They made viewing decisions after consulting listings, ads, critics' reviews, or word-of-mouth tips. Today, consumption appears to be radically different. An alphabet soup of technologies and business models has emerged: TVOD and EST (transactional video on demand and electronic sell-through, for example, Apple's iTunes), PVOD (premium video on demand, for instance, day-and-date services like Curzon Home Cinema), and FVOD/AVOD (free or ad-supported video on demand, such as YouTube), not to mention a vast pirate economy.[12] SVOD (subscription video on demand)—exemplified by streaming services such as Netflix, Amazon Prime Video, Hulu, MUBI, or the BFI Player—has become the fastest-growing means to watch films and series.[13] Of course, these models coexist with—rather than neatly replace—a panoply of largely thriving legacy media, terrestrial and satellite networks, digital and cable, rent-mailers, DVD kiosks, cinemas, a handful of leftover video shops, as well as other niche services.[14]

Rather than cinema schedules or TV listings, VOD services heavily depend on search and recommender systems to guide consumption among the often thousands of titles on offer. Personalized recommender systems for films and other cultural goods were whispered about at scientific conferences beginning in the 1990s.[15] Computer scientists define recommender systems as applications that "learn users' preferences in order to make recommendations" of cultural products. There are various recommender types. Some link the user to other users with similar consumption histories in order to forecast new items that might suit his or her taste. Others focus on the intrinsic qualities of the content itself to lead consumers to products they would not have discovered on their own. Still others perform a combination of these two methods. And yet all variants, according to their creators' technology- and novelty-driven accounts, require machine learning and software tools, deploying data (collected via user ratings or behaviors) and computational operations to predict and suggest appropriate consumption choices.[16]

Today, algorithm-led recommender systems are big business and constitute the core value proposition behind VOD services such as market leaders Netflix and Amazon Prime Video. One analyst reckons the Netflix homescreen—unique to each user based on his or her viewing history—to be the most powerful promotional tool in world entertainment.[17] Netflix proudly claims that its platform is free of "editorial content," in other words, devoid of human-generated evaluations or suggestions. The company boasts that its recommender system influences the choice for as much as 80 percent of its streamed content and is worth $1 billion annually in retained subscriptions and reduced marketing costs. Netflix chief content officer Ted Sarandos floated the story that the system's diagnosis of taste is so powerful that series such as *House of Cards* were "generated by algorithm."[18]

As such, Netflix's and Amazon's recommender systems represent the realization of the internet's early promises of personalization: content tailored to the *individual* user's needs, rather than broadcast to the wider public or identity-based demographic groups.[19] Rather than the opinions and insights of human experts, these recommendation engines bet on the wisdom of crowds: aggregations of passively communicated taste.

Less publicized amid the hype, however, is the continued existence of legacy recommendation forms. Word of mouth, an often intimately personalized means of suggestion, is practiced by Americans on average 16 times per day, including 2.8 instances involving media and entertainment products.[20] Box-office rundowns, top-ten lists, and best-of rankings are just a few examples that offer, in a non-personalized way, the same core feature as an algorithmic recommender system: a ranked list of items likely to interest the user. Likewise, niche VOD services, those with a cultural remit or aiming at a highbrow clientele (BFI Player, MUBI), use non-personalized recommendation styles that emphasize "human expertise" and "human curation," discovery and diversity. Their publicity downplays—or outright eschews—algorithmic suggestion. Rather than appealing to individuals' likes, they present films in ways that recall traditional forms of "good taste" and shared cultural norms of quality that date back to Gotthold Ephraim Lessing and Matthew Arnold: the hallmarks of many recommendation sources throughout the existence of audiovisual entertainment. They update such top-down conventions with appeals to community, social-media-esque sharing, and a contrarian artisanal identity in the era of big tech.

This book seeks to move the discussion of VOD *to* recommender systems, but also shift the discussion *of* recommender systems: to reposition where we locate and how we contextualize these phenomena. Almost all scholarly efforts on VOD recommender systems—not to mention treatments of these mechanisms in general, whether for music platforms like Spotify or Pandora, search engines like Google, or news feeds such as Facebook's—home in on algorithms and data collection.[21] Although such assessments vary somewhat in terms of moral valence (some celebratory and many others scathing), virtually all foreground quantitative computational processing as an ontological characteristic, the essence of what makes a recommender system worthy of its name. Indeed, these many articles and books propose the "novelty" of data, "datafication," the "datafied society," or "algorithmic culture" as the underlying justification for their study. They bracket algorithmic recommender systems as unprecedented in their features and unforeseeable (and, above all, unforeseeably wonderful or dangerous) in their effects on individuals and society at large.

Such theoretical pronouncements, however, remain unsatisfactory. Largely untested and sometimes speculative in nature, they are proving to be out of step with the available empirical evidence. How might we better understand algorithmic systems by *not* stipulating that their "revolutionary" features blow up long-standing user norms a priori? Might we better take stock of VOD recommendation not by considering them as sui generis technological forms but, rather, for their functions and uses and as continuations and transpositions of legacy forms such as word of mouth, criticism, and advertising? Indeed, how might we better understand these developments by reminding ourselves of the insights of media archaeologists who look for continuities alongside uniqueness, for the old in the new and for the novel already in the old?[22] These are questions that—in the hangover of new-media celebrations and lamentations—urgently require answers.

This book proposes to address these lacunae via two fundamental shifts of perception: from the assumption of novelty toward a historical genealogy, and from the prism of technological determinism toward a user-centric attention to the instrumental functions of these services. At this stage, I argue, it is more enlightening and productive to consider VOD recommender systems as much as of a piece with *Leonard Maltin's Movie Guide* and *Reader's Digest*, as with Google's search engine. What if we see Netflix's or Amazon's recommender systems within a subset that includes Rotten Tomatoes, Watching, MUBI's curation-style model, and distributors' posters—that is, on a spectral plane of promotion, suggestion, and information? These avenues of inquiry entail a detour from a synchronic fixation on the producer, on form as input, on (the myth of) big data, and on assumptions of (technological and thus functional) novelty. Instead, I submit, we must adopt a more instrumental and archaeological perspective that attends coequally to the user, to form as function, to novelty as discourse, and to diachronic pathways and overlaps.

The benefits of these shifts in approach deserve explanation. Previous assessments remain stubbornly producer-centric, lavishing attention on the engineering and business heroes or villains who create the algorithms and sell them as a service. From the perspective of the producer, it makes a significant difference whether a platform deploys an algorithmic recommender system or whether the interface arranges films and series A to Z with blurbs written by interns or professional critics: each of these decisions requires crucially distinct employees with different skill sets and labor costs, resolutely incompatible business models, and unique selling propositions (USPs). The data engineer is only interested in a recommender system that requires his or her talents and efforts to construct and maintain; otherwise it may as well not exist and will be certainly unworthy of the name. For the CEO of an online VOD platform whose economic value rests on matching content to individual users' unique prior viewing behaviors, an algorithm-led understanding of recommender systems is similarly logical.

From the perspective of *users*, however—who after all are seeking to watch content that interests, moves, occupies, or otherwise engages them—the narrow computer-science definition of a recommender system (a device that processes viewing data to learn preferences and algorithms to suggest content) may miss the mark. Whether or not a user discovers suitably engaging content by seeing a poster, scanning Rotten Tomatoes, soliciting a tip from a friend or stranger, or reading a title on a VOD platform homescreen is not inconsequential. (As we shall see, people react differently when asked to assess the trustworthiness of a recommendation they receive via a poster, a trailer, a critic's review, a best friend's tip, or an algorithm.) But, in general, the inputs of code on Netflix or Amazon ultimately remain—just like a critic's salary or a poster's ink—a matter of secondary interest. For the user, I will demonstrate, technological difference has some, but not overwhelming significance: for example, any given platform suggests a film with more or less description; with greater or fewer options for selection; with more or less need to consult Wikipedia, IMDb, a favorite critic, a friend, or further scrolling or searching. More basically, users evaluate the effectiveness and efficiency with which any given platform delivers content that they appreciate, a transaction that plays out each time they endeavor to watch a film or series. Over time, each mode of recommendation accrues a certain level of credibility and becomes subject to conscious and unconscious routines and rituals; for the consumer, the *outcome* and trustworthiness of recommendation, alongside the cost, are crucial. The empirical evidence presented in this book demonstrates that this semi-instrumental perspective obtains whether users scan TV listings (in print, on the screen, or via the internet), examine a film festival program, catch sight of a poster on public transportation, or view a pop-up ad: that is to say, in any and all film and series recommendation situations.

Approached from this vantage point, technology remains an important consideration, but neither determinative of any individual consumption outcome nor determinative of all consumption outcomes in aggregate. Using this framework does not represent mere contrarianism on my part. The empirical audience research on media choice presented here indicates that—*pace* both the new-media cheerleaders and the filter-bubble-thesis proponents—most consumers use (and prefer to use) traditional forms of information, especially word of mouth, much more often. Moreover, most people trust VOD recommendations much less than traditional information sources and suggestions. Algorithmic suggestions maintain some value to many VOD users, but they typically constitute just one small piece of a multistage, iterative process of active and passive engagement with film and series information.

In sum, this book approaches VOD recommender systems within a user-centric archaeology of cultural recommendation and media consumption choice. This is no fanciful enterprise. After all, when computer scientists were dreaming up algorithmic recommenders for films and music in the mid-1990s (themselves leisure-sector transpositions of so-called "expert systems," which, among other functions, helped doctors diagnose patients based on symptom inputs), they explicitly referred to legacy forms. "Collaborative filtering," the algorithmic modeling and prediction of user tastes based on similar users' viewing histories, deliberately sought to replicate video-store-clerk and close-friend word of mouth. Other programmers designed techniques and code with the experience of a trusted film critic in mind. Although this utility is forgotten, early algorithmic recommender systems emerged from data engineers' "simple observation," one computer science textbook reminds its readers, that "individuals often rely on recommendations provided by others in making routine, daily decisions" and commonly trust others' suggestions: peers' tips in selecting a book or a doctor, employers' letters of reference for job applicants, or critics' reviews in deciding what films to watch. Developers consciously sought to "mimic this behaviour," to transpose, scale, and automate these legacy forms, by applying computational processes to "leverage recommendations produced by a community of users and deliver these recommendations to an 'active' user."[23] Indeed, according to a biographer, Jeff Bezos's business model for Amazon depended on this remediation: in particular, of tips from experienced local shopkeepers, who knew their customers well enough to suggest to Customer X something in the vein of John Irving, and to Customer Y the next Toni Morrison.[24] Despite their supposed novelty, algorithms have existed for thousands of years as tools for humans to make predictions and prognoses. Some of the component tongue-twisting operations of Netflix's recommender system, such as Markov chains, have been a stock part of statistical modeling for well over a century.[25]

This book reveals that the conventional wisdom is wrong. Despite the by-now mainstream assertions that (algorithmic) recommender systems are decisive in forming opinions about what to watch (or indeed the more extreme contention that they may be brainwashing large swathes of the population), the studies that I have conducted and analyzed for this book suggest that while such mechanisms are hardly inconsequential, they still play a relatively minor role when considered among the myriad ways that we come to consume audiovisual content in the digital age. We must reckon with these services in a more differentiated and informed manner.

My perspective on this subject requires a more expansive definition of recommender systems, one that remains user situated and purpose based. This is necessary to correct computer scientists', media scholars', and marketers' primary focus on the specific technology of the application and the agency of the commissioning producer, developer, or programmer. I see recommender systems within an algorithmic and non-algorithmic spectrum of methods and applications to guide consumers' selection of cultural products (and here especially films and series). Recommender systems—regardless of whether they are employed by Amazon, BFI Player, Hulu, MUBI, Netflix, or another provider—rely on more than programmers' algorithms to achieve their aim. They must have an interface and therefore a particular layout and design. This interface may span (and vary) across multiple devices and may include multiple media and delivery systems (e.g., emails or push notifications with suggestions of titles likely to interest users). Conceptually, recommender systems must present a choice architecture in categories, lists, rankings, and sequences. In order to be coherent they must both display and hide content, foreground and circumscribe it; they must somehow inform, describe, contextualize, compare, or otherwise represent content choices. And they must perform these functions in combination with larger demands and limitations, such as finite acquisitions budgets and geographical and temporal constraints on exhibition. Adopting this more functional and user-centric definition allows us to consider various VOD providers—even MUBI or BFI Player, which eschew algorithmic suggestion in their rhetoric and technological forms—as deploying, in effect, recommender systems. It permits us to forgo bean-counting Markov chains and hair-splitting Bayesian networks in order to arrive at first principles: that in fact *all* of these services seek to promote certain films or series to certain subsets of users in order to provide a manageable and compelling content choice. It remains necessary to account for those variables with nuance, rather than falling prey to a tribalism against numbers, fallacies of novelty, and other forms of disciplinary border policing and PR claptrap.

This book considers VOD recommender systems and the cultural phenomena that they seek to simulate, complement, and supersede. Although algorithmic recommender systems are now used widely across portals of audiovisual content—not to mention travel websites, financial services, and medical diagnostics—this book concentrates squarely on SVOD services such as Netflix, Amazon, or MUBI rather than AVOD (e.g., YouTube) or TVOD/EST (e.g., iTunes) platforms for a number of reasons. First and foremost, SVOD represents the fastest-growing means of viewing films and series—despite the efforts of studios to follow on the richly rewarding experience of VHS, DVD, and Blu-ray in the form of TVOD/EST digital files. Second, SVOD employs not only a unique business model, but also offers an essentially different user experience: the oft-changing variety of films available on any given day means that a sophisticated form of recommendation must be built into the overall service. In essence, TVOD/EST represents a nonphysical form of DVD/Blu-ray, whereas SVOD follows on the economic structure and consumer experience of pay television. Third, TVOD's usually large catalogs (compared to SVOD) and AVOD's open-ended nature (YouTube users are uploading thousands of new videos per day) demand different discoverability and recommendation logics and logistics. Fourth, although YouTube and its competitors are significant social phenomena, contain some feature films and series, and use algorithmic recommender systems, overall they represent a parallel industry of amateur and user-generated content, and, as such, a distinct issue not easily integrated into the focused discussion of a single monograph. Fifth and finally, SVOD's means of suggestion—on first gloss at least—represent the greatest challenge to traditional forms of criticism and other traditional attention focalizers, eliciting the most substantial concern in the "new algorithmic culture" discourse that I seek to interrogate.[26]

In these exhilarating days of media change, digitally induced proliferations of selection, and perceptions of information overload, recommender systems have emerged to manage attention and to organize choice in films and series. This book seeks to account for SVOD's means of recommending audiovisual cultural goods and its promises of connecting content to users in a putatively new way, how real users respond to recommender systems, and what this may mean for society at large. Investigating above all the world's most-subscribed VOD service, Netflix, this book dissects the marketing rhetoric, industry talk, cultural logics, technical processes, business models, historical transformations, user credibility, and social ramifications of recommender systems.

These objects of inquiry stem from and inform my main research questions. First, I ask, why do recommender systems exist and what social needs do they purport to satisfy? How do VOD services (and, in particular, Netflix)—in their business models, interfaces, and marketing rhetoric—propose to connect their users to audiovisual content? Second, how do they seek to explain recommendations and assure users of the credibility of their suggestions? How do they purport to individualize, personalize, connect people, or foster communities? To what extent are their choice architectures, their selection-and-filtering methods and mechanisms, transparent to, or hidden from, the user? Third, to what extent are these developments truly new? How do they—in their design and function—copy, challenge, transpose, substitute for, or replace preexisting or historical forms of film and series information, evaluation, or recommendation, such as criticism, word of mouth, and advertising? Fourth and finally, to what extent and how do *real* users deploy VOD recommenders? How do they understand, rate, rank, conceptualize, imagine, and talk about these systems? To what extent are these systems *actually used* in conjunction with, or instead of, legacy forms of cultural recommendation?

Responding to these lines of investigation and subjecting these systems and their real-life use to empirical and historical scrutiny will show that neither the utopian nor the apocalyptic modes of commentary and scholarly analysis—the prevailing discourses emerging from the business and tech worlds, from journalists and academics—is justified. Human beings maintain innate desires to choose and consume audiovisual storytelling that occupies them, moves them, enlightens them. Over the years, ensembles and routines of film and series suggestion and evaluation arose to address this need. That recently canny businesspeople have developed systems that use web interfaces, computational applications and processes, and algorithmic formulas to perform the tasks that listings, guidebooks, critics, posters, video clerks, or friends (used to) undertake must be studied closely. But the novelty of numbers cannot obscure the fact that such systems are just the latest means, in a broader arc of media history, to execute a long-needed function; the rise of algorithmic recommender systems cannot foreclose the reality that other solutions to this problem have not disappeared. Beyond the obvious advances in computer science, and despite sustained marketing rhetoric and equally fierce criticism, recommender systems are neither as revolutionary nor as alarming as their celebrants and critics, respectively, maintain.

Rather than unique operators that animate hitherto unprecedented technologies and cultural logics, VOD recommender systems emerge out of the past, retaining and reconfiguring the purposes and affordances of both legacy and long-forgotten recommendation forms. This book will illustrate how today's platforms appropriate traditional functions of interpersonal communication, cultural mediation, humanistic criticism, and arts education. And yet, I shall argue that VOD recommender systems remediate—rather than replace outright—the cultural gatekeeping techniques of word-of-mouth communication, critical reviews, and advertising. They intervene into the ecology and economy of attention and provide new accents and velocities, rather than radically transforming taste, culture, and humanistic labor. The evidence will show that VOD recommender systems diversify, rather than displace, legacy forms. Examining real users' experiences reveals that VOD recommender systems are most often deployed in addition to, rather than instead of, an ensemble of legacy recommendation sources.

Important aspects of VOD recommendation are not only not new: they are not nearly as widely used or as widely trusted as assumed. My studies of real audience reactions to various recommendation sources reveal how these systems attempt to accrue, but largely fail to attain, users' trust. To be sure, VOD recommender systems offer important ways to attenuate perceptions of too much choice and simplify decision-making in the digital era of cultural plenty. Nevertheless, the evidence will show that their use is hardly widespread let alone universal. Their credibility, among wide swathes of the populace, is exceptionally low compared to almost all other forms of recommendation, including advertising. Despite pervasive assumptions on both sides of the debate, consumers are not blind to the new forms, nor do they blindly trust them. Yes, VOD recommender systems represent significant advances in computer science and coding; the sophistication of their design is not in doubt. Nevertheless, this book will show, technological complexity is a poor predictor of user trust or of a satisfactory recommendation outcome. Legacy forms, such as "simple" word of mouth, enjoy more credibility, by unmistakably substantial margins.

Given these facts, this book will conclude that the theories and prophecies of AI's wholesale hostile takeover of humanistic culture cannot be sustained, at least for the medium-term future. Rather, the preponderance of evidence intimates a more or less unpeaceful coexistence: on the one hand, a supersession of some low-level forms and formats of information and suggestion that algorithmic systems can more efficiently perform, disseminate, and scale, but on the other, an overall persistence and even increased need for value-adding human cultural mediation and evaluation. Although scholars have often celebrated the new media as ushering in the democratic "death of the gatekeeper," I will demonstrate that, by some significant measures, the thirst for filters, curators, and critics is stronger than ever. In sum, this book challenges widespread assumptions about the effects of algorithmic computational processes and big data on media choice, revealing that there may be more continuity than change in the digital age.

NEW-MEDIA TALK AND THE TECHNO-DETERMINISM TRAP

Answering the multidisciplinary questions at the core of this project requires similarly complex and comprehensive methods, ways of acquiring knowledge that eschew territorial disputes between fields and disciplines. In order to amass and analyze the data that would allow me to best understand recommender systems and their role in film and series choice, I proceeded with a methodology aligned with the goals and procedures of what has been called critical media industry studies.[27] My approach encompasses three main thrusts: (1) business and technology history, in particular, the history of Netflix; (2) formal analysis of these systems and comparison with legacy recommendation sources; (3) analysis of empirical audience studies conducted by my research team as well as those submitted by others over the years.

Clearly, understanding VOD recommender systems necessitates close formal and content analysis of interface design and programming models, websites, and applications—both in the "today"-time of research as well as archived copies. Expanding the purview from narrow cubbyholes of technology or text alone, however, required the dissection of further overlapping and interlocking concerns: business strategies and promotional rhetoric, media ownership and organization of labor, consumption practices and cultural commentary, not to mention theories of media change. In order to undertake such an analysis I pored through archives, examined myriad (film, business, technology) trade papers and computer science secondary literature, assessed industry reports and government policy documents, and anatomized press releases and interviews with various CEOs and their subordinates. Supplementing these sources, I conducted correspondence and interviews with employees of these companies and with their press officers, aware of my role as intervening agent and careful to treat these sources— like all others—not as transparent reflections of some immutable truth, but rather as symptoms of industry talk and self-explanation, perspectival lore that, after subsequent analysis, became instrumental to understanding larger constellations of power.

In my endeavor to understand audience behaviors, I also needed to conduct my own empirical studies of user reactions to recommender systems in the context of legacy sources. This included commissioning and quantitatively analyzing representative national surveys in the United States and United Kingdom. Furthermore, deepening these insights required my research team to conduct several dozen individual user interviews, which I subsequently dissected qualitatively to more fully understand how viewers use recommendation sources and arrive at film and series choices. All of this data needed to be reconciled and triangulated with a wealthy history of empirical findings into how people respond to different forms and instances of cultural recommendation, studies that have emerged from media audience research, cultural and behavioral economics, social psychology, sociology of art and taste, marketing, and computer science.

In conceiving the parameters of this project, it was important to tread carefully around the most common trap of prior approaches: technological determinism. Historians have demonstrated that new media technologies—whether the telegraph, telephone, cinema, or television—rarely have the immediate and most dramatic effects predicted at their introduction.[28] And yet too often when dealing with phenomena associated with today's new media, observers neatly ascribe grandiose visions of utopian horizons or social enervation to technological forms alone. Rather than drawing permanent and intrinsic lines of causation from technology to social consequences, we need to allow for more tenuous correlations (how social and technical processes mutually codetermine each other) and be aware of the deeper ecologies that link technology, organization, and discourse. As a number of scholars have emphasized, we must see new media (and their technological delivery systems) within the context of their social settings as well as users' engagements and motives. Any medium, the best observers maintain, must be understood not just for its logistics of communication, but for what Lisa Gitelman calls its associated "protocols": the behaviors, language, and other social practices that emerge and evolve around the medium's use.[29] In this way, the book engages an archaeology of *film and series culture and choice mechanisms*. Cultural intermediaries, I submit, deserve scrutiny equal to, and as rigorous as, that long devoted to delivery systems and audiovisual content.

Attending to social contexts calls for parsing the language of new media and subjecting it to more than a modicum of critical analysis. Over my career I have become ever more convinced that language (including audiovisual rhetoric, such as posters or trailers) offers some of the most valuable insights into the inner workings of cultural phenomena. Words matter. How we describe, evaluate, and position artistic and economic products (and most films and series are best considered as a hybrid of the two) reveals intentions, agendas, and values. A "shithole" and a "country with economic and social problems" may be functionally equivalent synonyms, but they imply vastly different perspectives and moral judgments. In order to understand digital technology, we need to understand how people try to explain, position, justify, convince, sell, and moralize—to consumers (B2C) but also internally to each other within the industry (B2B). For the present project, this task required an ensemble of sources, including annual reports, industry conferences, press releases, mission statements, company newsletters, interviews, and advertising. These materials help us understand how VOD companies pitch recommender systems to customers, rationalize their design and economic value, and, overall, construct a social reality.

The revolutionary rhetoric (and doomsday-mongering) surrounding the economic and sociocultural potential of the internet and digital technologies is by now a well-established phenomenon. Articles and books make bold and often aphoristic predictions about how the new media will fundamentally alter human values and relationships, transforming our experiences of space and time and ushering in reinvigorated social hierarchies and political systems. Indeed, a number of academics have illuminated how the language of new media—today's but also yesterdays' new media—constitutes a "vapor theory," a confluence of geek jargon, business shoptalk, and slick marketing palaver.[30] It behooves us to think critically about the mismatched incentives of those who continue to recycle this pap.

In particular, Vincent Mosco's *The Digital Sublime* takes the prognosticators and celebrators of the new media to task by contextualizing the discourses surrounding the internet and big-data computer processing within the longer history of older "new media," such as the telephone, radio, and cable television. His study offers an important primer on how to approach the grand visions ascribed to recommender systems and matters of choice in audiovisual culture. Mosco shows that new media are almost always greeted with a revolutionary rhetoric that promises wide-reaching effects on society. Before the internet and on-demand viewing, commentators claimed that cable television had a power to connect people in a way never experienced before, that a new and all-reaching bidirectional form of multichannel communication would arise and "usher in a Wired Society governed by Electronic Democracy" that would "revitalize communities, enrich schools, end poverty, eliminate the need for everything from banks to shopping malls, and reduce dependence on the automobile. If we only had the will, the money, the right policies, etc., etc. In short, cable TV would transform the world."[31] Similar pronouncements welcomed the telegraph, which would supposedly erase borders, end warfare, and set up "the kingdom of peace" (120), and electrification, which was predicted to end all crime. The telephone, according to contemporary observers, would prove to be a "business savior" that would set up distance shopping, provide housewives with liberating freedom, ensure family safety, end the necessity of writing and thus help illiterates, and otherwise create a "new social order" (126). Just as in the last twenty-five years a set of gurus and promoters (Nicholas Negroponte, Al Gore, Steve Jobs, or Mark Zuckerberg) have declared a digital, internet, or computer age, so too did any number of marketers, journalists, academics, and politicians speak of the Telegraph Age, the Age of Electricity, or the Radio Age (2, 20, 117–18). Using seemingly compelling but ultimately utopian and fruitless metaphors, these "cosmic thinkers" endow new technologies with sacred and sublime powers that fulfill long-standing and primal desires and needs. Digital libraries that house all existing information and remain instantly accessible. Electronic commerce traveling across information superhighways that radically simplify logistics and eliminate distribution bottlenecks. Ecologies of digital information flows yielding organic virtual communities that develop social experience and promote harmony. With insistent suggestions that humans have never experienced anything comparable to these developments, such commentators build narratives that are supposedly inevitable and unchangeable, not to mention unprecedented and therefore immune to historical comparison (37, 51–52, 82). In this way, Mosco argues, "the myth encourages us to ignore history because cyberspace is genuinely something new, indeed, the product of a rupture in history, the Information Age. Until now, information was scarce; it is now abundant. Until now, communication technology was limited; it is now universally available at prices that are rapidly declining" (34–35). The denial of history via the rhetoric of the unprecedented, central to the logic of the myth, evacuates the possibility of alternative scenarios and human agency.

Mosco's project, a worthy one for all histories of new media, aims not to simply dispel or refute these myths: for example, that the internet and computers will revolutionize society by connecting people across the world, giving rise to an information-, entertainment-, and knowledge-based economy, or that we are experiencing a decisive communications and media transformation that will help us transcend traditional configurations of time, space, and power structures (18, 2). Instead, the *social functions* of such new-media myths are crucial for their substantial symptomatic value, "stories that animate individuals and societies by providing paths to transcendence that lift people out of the banality of everyday life" (3). Indeed, following Mosco, it seems that the myths surrounding digital culture "are important both for what they reveal (including a genuine desire for community and democracy) and for what they conceal (including the growing concentration of communication power in a handful of transnational media businesses)" (19). In this sense we should read the (with historical retrospection downright fanciful) forecasts about the telephone or electrification less as marketing fraud and more for the social desires they seek to innervate: a connected, more harmonious world without racial strife.[32]

It will be a key task of this book to uncover and decode the hopes and dreams—but also nightmares—that lie behind recommender systems and matters of cultural mediation and choice in audiovisual culture. Fantasies of instantly sating tastes with minimal effort. Of bespoke, personalized suggestions tailored to momentary moods and derived from the wisdom of crowds, rather than the broadcasted whims of elites. Empire-building fantasies about completeness and mastering whole stores of information and categories: directors' oeuvres, complete series and episodes, far-flung national cinemas, plus new, hitherto unprecedented and even "alien" categories. The freedom from imperfect prints, unreliable analog television reception, decaying video-cassettes, scratched DVDs, malfunctioning TiVos, exorbitant cinema tickets, long trips to video stores and faraway theaters, and bloodcurdling cable bills. A democratic vision of equal access and empowered individual consumers among invigorated networks of viewers and communities of interest. The fundamental belief in, and alternately fear of, the new. As we approach VOD recommender systems, their promised innovations, and their dreaded consequences, let us bear these insights in mind. For the visions and promises of VOD platforms and recommender systems are symptomatic in Mosco's sense: they reveal broader social desires for personalization, democracy, progress and innovation, community and social cohesion, diversity and simplicity, and time-saving, cake-and-eat-it ease of use.

NETFLIX RECOMMENDS

Looking ahead to anticipate our path, the first two chapters examine how VOD recommender systems intervene into ecologies of cultural mediation, taste, and choice. Chapter 1 explains the social needs and consumerist draws behind recommender systems: to focalize attention and reduce perceptions of "too much," allowing users to feel they have come to a satisfying choice. Although recommendations have existed almost as long as culture has been produced, an economy of attention exists in the contemporary audiovisual landscape for two key reasons: first, because of "nobody-knows" qualities intrinsic to films and series as experience goods; and second, because of the rapidly expanding quantities and availability of content in the digital age. Furthermore, this chapter puts VOD recommender systems into dialogue with long-standing research on the uses of legacy suggestion sources: especially criticism, word of mouth, and advertising. In turn, chapter 2 explains how VOD recommender systems work conceptually, by introducing the spectrum of algorithmic and curation-style services, anatomizing their differences by drawing distinctions between the two in terms of technological form and interface, business model, marketing rhetoric and attitudes toward personalization, mediation of taste, modes of credibility building and explanation, and scalability.

Chapters 3 and 4 bear down on a close case-study analysis of Netflix. The market-leading VOD service features perhaps the world's most recognized recommender system, one that stands in for a whole array of competitors, such as Amazon. Indeed, Netflix's recommender system represents the business's USP and self-declared most precious asset. Although both the company and its critics claim that the algorithmic recommender system is unprecedented and unique, the origins and development, under scrutiny in chapter 3, demonstrate how company engineers consciously remediated video-store classifiers, video-clerk suggestions, broadcast criticism, top-ten lists, peer word-of-mouth tips, and other legacy recommendation forms. Netflix's persistent rhetorical focus on personalization and its ideological conviction to rid itself of "editorial content" have long functioned as key aspects of its branding and credibility efforts—and targets at which critics have long aimed. Nevertheless, this chapter shows how the recommendation engine is optimized to blend personalization with factors such as critical acclaim, short- and long-term popularity, novelty, and diversity. Despite the focus on revolutionary or at least novel practices, the chapter shows, Netflix's recommendation styles and designs derive largely from past forms and norms.

In turn, chapter 4 illustrates how Netflix builds its credibility—and inspires concern among media scholars—with PR appeals to scientific precision and objectivity. As the company transitioned from a DVD-by-post operation into a fully online streaming service, and its recommendations downgraded users' self-selected stars in favor of surveilling behavior, it purposefully shrouded itself in a myth of big data, eagerly garnering press accounts of an "alien," "superhuman" quality to its operations and capabilities. This chapter demonstrates how, on the one hand, Netflix's performance of scientific objectivity, innovation, and differentiation constitutes a credibility-raising exercise akin to how critics have traditionally performed their authority, knowledge, and distance to (or familiarity with) the industry in order to establish their trustworthiness. On the other hand, the discussion suggests, despite this mythmaking, the user experience remains decidedly different.

Chapter 5 makes real users and uses of recommender systems its direct object of inquiry. It deploys the theoretical framework of "folk theories" within a mixed-method empirical audience study, including two nationwide (United States and United Kingdom) representative surveys and several dozen in-depth interviews. Overall, the chapter seeks to understand the extent to which recommender systems actually function as a primary means to select and access films and series; users' understanding of the algorithmic mechanics; how they speak about the effectiveness of, and their trust in, recommender systems vis-à-vis other sources of information; and how they in general talk about their use of recommender systems. The analysis offers a decidedly mixed picture of how real people engage with recommender systems. First, recommender systems are not a primary reference for most users; rather, some deploy them in a multistage process of consulting various information sources. Second, most users, although hardly experts on the technical specificities, have at least a functional, often selective understanding of how algorithmic recommenders work in practice. Third, although some users absolutely avoid or devotedly follow the suggestions of recommender systems, many use folk theories to rationalize their ambivalence toward them, including persistent ideas that the tools fail to decipher their moods or that no algorithm could ever understand eclectic human tastes. Fourth and finally, the chapter wraps up with an inventory of user typologies and discursive themes that emerged inductively from the interviews, modes and manners of contemporary film and series choice that indicate the altogether varied and idiosyncratic uses and understandings of recommender systems. This analysis provides nuance to both utopian and dystopian narratives of these devices' effects.

Chapter 5 yields to an afterword, which carries this discussion to larger humanistic concerns regarding the automation and quantification of culture and taste. The afterword sums up the data and analysis to suggest that— despite the undisputed sophistication of the technology—neither Netflix nor its chief competitors offer users a compelling experience of recommendation akin to criticism beyond capsule reviews or even word-of-mouth episodes. VOD recommender systems hold sway above all in low-stakes viewing situations.

ONE

Why We Need Film
and Series Suggestions

WHY DO RECOMMENDER SYSTEMS EXIST in audiovisual culture? What social needs do they seek to satisfy? What are their fundamental technological forms and their providers' key business models and cultural promises? To what extent are they novel, and how do they borrow from legacy recommendation forms? These are the central questions that chapter 1 and chapter 2 begin to answer. The following pages begin by demonstrating how, in general, recommendation mitigates a basic problem of films and series: their status as experiential consumer products. I then go on to contextualize VOD recommender systems within a longer history of information regimes that seek to focus consumers' attention and direct their decision-making. The second part of the chapter shows how recommendations—and recommender systems—have become especially necessary in a digital age in which barriers to production are low and content proliferates. Overall, I argue that we should recognize VOD recommender systems as mitigators of risk and agents of surplus. There has always been a need for gatekeepers, filters, attention focalizers (including promotions and publicity), curators, and recommenders, because overproduction is the "rational organizational response" of the entertainment industries' "nobody-knows" principle.[1] Algorithmic VOD recommender systems, as a part of a larger spectrum of cultural recommendation sources, speed up, personalize, automate, and scale the suggestion of audiovisual goods, for which taste varies greatly from user to user. Arbiters of qualitative risk and quantitative excess, they base their tips on putatively objective aggregations of user behaviors, rather than the pronouncements of human experts.

Despite contentions that algorithmic recommender systems are fundamentally novel, we can learn much about their basic social functions and instrumental uses by examining the long tradition of research that has been conducted on legacy cultural recommendation forms. Indeed, a whole slew of theory and empirical findings in film and media studies, sociology of art, marketing, and other disciplines has sought to account for the role of suggestions in selecting cultural goods. A long strand of social psychology and cultural economics dictates that because every film and series is essentially unique and because taste for audiovisual products is particularly fickle, recommendations remain crucial.

This distinction becomes clear when comparing films and series to other major types of consumer products. In general, *search goods* (such as laundry detergent or cameras or cars, etc.) are those that can be easily evaluated before purchase and essentially substituted by brand or model.[2] Every bottle of Persil (or Tide or any other brand) is identical and consumers put themselves in little economic jeopardy by sampling one of the relatively few varieties or another. In contrast, audiovisual storytelling is by most measures an *experience good*: each and every film or series is a prototype without a reliable indication of its capacity to please its viewer. The ability to predict quality in advance of consumption, in other words, is notoriously difficult; theoretically, each exemplar must be consumed in order to be properly judged. Of course, media producers attempt to reduce risk with reboots, sequels, adaptations of familiar and proven intellectual property, or by employing staff whose track records demonstrate success in satisfying viewers. (There is a reason, after all, that they resurrect *Avengers* and *Star Wars* and *Batman* year after year, whereas first-time filmmakers must beg, usually unsuccessfully, for production funds.) And yet they must build an audience for each new title. Scholars speak of a "symmetrical ignorance": neither producer nor consumer, neither seller nor buyer can reliably assess the success of the product before the transaction and reception.[3]

Furthermore, films and series pose special risks because consumption motivations and evaluation criteria can differ wildly from consumer to consumer. To wit, almost all consumers use laundry detergent for their washing and will evaluate it as effective if their clothes smell, feel, and look fresh after use. In stark contrast, films and series are used to entertain, educate, fantasize, pass the time, socialize with others, inhabit different perspectives, explore new cultures, and for a whole host of other reasons.[4] Media industries, as Janet Wasko and Eileen R. Meehan write, "produce commodities that convey narratives, arguments, visions, symbolic worlds, and imagined possibilities."[5] Whereas differentiation between two bottles of the same laundry detergent would be grounds for concern and complaint (and many people use the same brand and type week after week, year after year), consumers expect and enjoy originality and novelty across films and series, which many will only view once.[6] The prevailing uniqueness, however, results in a fundamental consumer ignorance regarding the quality of the product that they have not yet consumed. "The uniqueness, which cultural goods must demonstrate according to convention," Joëlle Farchy states, "leads to *uncertainty* about their quality, which in turn unsettles the consumers' traditional selection processes. The appropriate means to limit this uncertainty is for the consumer to acquire *information*."[7]

This information has traditionally inhabited three general forms across three different sets of recommendation sources: first, advertisements (including posters, trailers, television and online spots); second, the interventions of public intermediaries (especially experts, critics, and journalists) in the form of essays, reviews, interviews, and other contributions; and, third, peer word of mouth, including overheard comments and direct tips from friends, family, acquaintances, and strangers.[8] Each of these information sources and means of suggestion will possess varying amounts of *credibility* with consumers, who will negotiate between one or multiple forms of recommendation based on their preferences in general or in relation to the individual content and potential portal (cinema, television, VOD, and so on). For example, some studies differentiate between "lay" and "expert" reviewers, determining that while consumers more readily accept the opinions found in lay reviews of search goods (like household appliances), they tend to be more skeptical of non-professional online reviews of experience goods such as films or music albums.[9] Other studies show how, in general, word of mouth trumps critical reviews in terms of box-office influence.[10]

Moreover, individual viewers will seek out and use recommendation sources differently according to their level of media consumption, as well as their personality. Numerous audience studies, conducted from the 1950s until as recently as 2014 in a variety of European countries and in North America, have repeatedly demonstrated a statistically significant difference in the attention paid to various types of information sources according to frequent and occasional users.[11] Those who consume fewer films and series valued television information (television ads, coverage, publicity) and above all word of mouth in deciding whether to see a film and attended much less to considerations such as critics' reviews, newspaper coverage, festival prizes, and the name of the director. In contrast, heavy users—a decided minority (roughly 10–25%), namely, frequent cinemagoers, TV junkies, festival attendees, and cinephiles—maintained the opposite habits, tending to engage more with production news and critical reviews.[12] These results are confirmed in a plethora of other studies over the years, which suggest that there are fundamental distinctions to how various sorts of consumers respond to different kinds of recommendations.[13] Still further studies have revealed differences in how consumers react to positive vs. negative recommendations. Certain types of viewers, especially light users, are more likely to be influenced by negative recommendations—in particular, negative word of mouth—whereas heavy users ignore word of mouth and tend to follow their own initial impressions and judgments, deferring only to select, trusted critics.[14] In turn, social psychologists and behavioral economists have discerned how personality types and choice heuristics further inflect these truisms about cultural recommendation: media users with a "maximizer" personality will scour long lists of films and series, and multiple recommendation forms, to find appropriate content. In contrast, "satisficers" make little effort when choosing, because they are more apt to be content with adequate, rather than perfect, selections.[15]

Furthermore, the evidence overwhelmingly suggests that the precise role of recommendations will differ according to source and also according to the type, experience, and sophistication of the audience. These conclusions already begin to cast some doubt on the more extreme filter-bubble arguments, which imply that indecipherable, and thereby pernicious, algorithms uniformly lead the blind masses astray. They beg the question of whether light consumers of audiovisual storytelling will value recommender systems more highly and follow them more closely. In this context it is also important to consider another observation made in several areas of the research: that those consumers who undergo a "learning process"—not only about the particular exemplar but regarding similar products (e.g., other films and series; other films and series of that particular genre, type, or style; or other works by the creators of the product in question)—will be better able to predict the quality of the exemplar before consumption.[16] Indeed, computer scientists envision recommender systems to have different uses for consumers according to their level of media literacy, and aim them primarily "toward individuals who lack the sufficient personal experience or competence in order to evaluate the potentially overwhelming number of alternative items" on offer.[17] It is intriguing to speculate on how recommender systems mimic and systematize this learning process, possessing a well of knowledge (i.e., huge stores of data) to which otherwise only the learned film critic or a culture-vulture friend would be privy. The purposes and uses of recommendation could thus differ according to where a user stands in the learning process: although more sophisticated or frequent consumers may be more heavily engaged in media consumption preferences and thus more attentive to recommendation forms, light users may equally be engaged in intensive information searches, because of their general caution and selectivity in consumption.[18] We will begin to see more concrete answers to these questions in chapter 5, which tests these theories with an empirical audience study.

To sum up a mountain of research over decades: recommendations have a fundamental, constitutive, valuable, and unique role in the consumption of films and series, precisely because of the special qualities of audiovisual products. A consumer may decide on a laundry detergent brand once or vary his or her choice each month according to the price; he or she may take great pains to decide which camera or car to buy but will only purchase such an item once every several years. In contrast, films and series maintain vastly more variety and most people consume them much more frequently. Recommendations help us reduce the considerable uncertainty related to the quality of cultural products before their consumption; they also reduce the considerable effort that we would otherwise need to research, and thus better judge, the good.[19] Indeed, there is evidence that in early years computer programmers considered these opportunity costs for item searches—the stuff of research behind critics, advertising, and other recommendation instances—while assembling their applications.[20]

Some of the preceding scholarship was conducted in the pre-digital age, a time when start-ready films and series streaming over the internet seemed like science fiction, a futuristic luxury worth much more than $10 per month. When Farchy was writing in the early 1990s, the functions, roles, institutions, means of connection with, and social attitudes toward "experts and critics" were different. The local or national newspaper, specialist magazine, or trusted guidebook has yielded to unprecedented access to a whole host—some would say *glut*—of (professional and amateur) critics and reviews of various quality online.[21] An information-poor environment in which we might have assessed eating at a restaurant in an unfamiliar city by such primitive measurements as the number or brand of cars in the parking lot has given way to an overload of readily available aggregated opinions from sometimes thousands of prior users on TripAdvisor, Yelp, or any of the dozens of similar apps. The old "lemon problem"—the asymmetry of information between buyer and seller by which the latter could easily pass off a dud used car on the former—has been replaced by a deluge of available indicators from which to pick and choose. In turn, the sources of "word of mouth"—traditionally conceived as "informal communications between private parties concerning evaluations of goods and services"—has transformed from the recommendations of trusted friends, local acquaintances, and the neighborhood video clerk to a panoply of YouTube influencers, Twitterati, and other human sources from across the globe.[22] Although the ubiquity of online punditry and evaluations for various products has, by some measures, seemed to dampen the credibility of recommendation overall, one study suggests that most people remain remarkably receptive: 82 percent of American adults say they read online reviews; negative reviews have been shown to especially affect viewing behaviors.[23]

Indeed, despite developments in technology, matters of trust in the recommendation source remain as central as ever. Rather than the much-discussed issues surrounding algorithms and data, credibility constitutes the most important element in the effectiveness and success of recommender systems. Both the vast literature on persuasion and computer scientists' more recent evaluations of recommender systems demonstrate that humans are more likely to accept suggestions from trustworthy sources.[24] How and to what extent VOD services seek to enhance the credibility of their recommender systems—and real users' acceptance or disavowals of these appeals to trust—will feature as an important issue throughout this book.

CULTURAL SURPLUS AND THE ATTENTION ECONOMY

It is not only the particular quality of films and series as risky experience goods that creates a demand for recommender systems; the sheer quantity of choice similarly incentivizes them. In general, the more products on offer, the more any given consumer will need to rely on recommendations in order to arrive at a selection.

In the last twenty years, there has been an explosion of media production and consumption choices. Furthermore, dissemination platforms—from

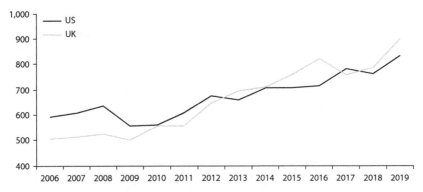

FIGURE 1. Number of films theatrically released per annum by territory. Sources: Motion Picture Association of America, *MPAA Theatrical Market Statistics 2015* (Los Angeles, CA: MPAA, 2016), 19; Motion Picture Association of America, *MPAA Theatrical Market Statistics 2016* (Los Angeles, CA: MPAA, 2016), 21; Motion Picture Association of America, *2018 THEME Report* (Los Angeles, CA: MPAA, 2019), 17; Motion Picture Association, *THEME Report 2019* (Washington, DC: MPA, 2020), 20; British Film Institute, *BFI Statistical Yearbook 2011* (London: British Film Institute, 2011), 12; British Film Institute, *BFI Statistical Yearbook 2016* (London: British Film Institute, 2016), 17; British Film Institute, *BFI Statistical Yearbook 2017* (London: British Film Institute, 2017), 7; British Film Institute, *BFI Statistical Yearbook 2018* (London: British Film Institute, 2018), 14; British Film Institute, *BFI Film at the Cinema* (London: British Film Institute, 2019), 10; Tim Dams, "U.K. Cinema Admissions Hold Up in 2019 Despite Streaming Boom," *Variety*, 16 January 2020, https://variety.com/2020/film/global/uk-cinema-admissions-cinema-first-2019-1203469103/.

cinema screens and television sets to film festivals and VOD platform usage—have increased significantly. Nevertheless, the number of hours spent on watching series and films has stagnated. This means that consumers are having to become more selective in their decision-making. Some statistics illustrate these points in a graphic way.

Since the middle of the twenty-first century's first decade, there has been a steady upturn of films in theatrical distribution. Although some years experienced declines along the way, in particular in the aftermath of the 2008 finance market collapse and the 2020 COVID-19 pandemic, the general trend has been an overall increase: 41 percent in the United States and 78 percent in the United Kingdom between 2006 and 2019 (Figure 1). Examining total films *produced* per annum (rather than those in theatrical distribution) gestures to an even starker perception of overload. The number of European films exploded in the ten years between 2007 (1,444 features) and 2016 (2,124 features), totaling a staggering 18,000 across the period.[25]

Industry insiders warn of an unsustainable production economy in North America and Europe, where neither the mass media nor newspapers nor niche online outlets can cover all releases—let alone that lay audiences could ever hope to watch more than a small sliver.[26]

Of course, the competition for audiences is not limited to feature films. Many would argue that the surge and quality of "television" drama—that is, series—has been the most important driver in content production. Indeed, the data bear out the fact that the growth in series has outpaced even feature filmmaking.

Table 1 and Figure 2 show how the number of original scripted series in the United States has more than doubled since 2009, with annual growth averaging nearly 10 percent. This figure (including the reference data from 2002) also demonstrates how online SVOD services (chief among them Netflix and Amazon) are driving the bulk of this growth. These platforms now produce more than a third of original scripted series and issue regular press releases about the billions of dollars they are investing in original programming, chiefly series. In 2017, Netflix spent $8.9 billion on content; Amazon ($4.5 billion) and Hulu ($2.5 billion) were ramping up their own budgets in order to catch up. Netflix alone has boasted of producing seventy-one series in 2017, a tally that does not even include its non-English-language and children's series.

There is no sign that the massive buys will abate soon. Indeed, Netflix reportedly expended $12 billion on acquisitions in 2018—including 82 original feature films and 700 new or exclusively licensed series—and approximately $15 billion across 2019. One market research firm predicted that by 2022, Netflix, Amazon, and Hulu would be paying up to $20 billion annually for original programming alone.[27] Even though overall viewing time for online moving images is stagnating, subscription-based streaming platform use—above all Netflix and Amazon—has surged year-on-year.[28] And yet, in a fragmented marketplace, the top 20 shows reach only 10.8 million domestic viewers, an equation that many in the industry deem insufficient to sustain the expensive proliferation of productions.[29] In 2016, US subscribers to Netflix (4,563 films; 2,445 series) or even the average SVOD service (1,285 films) faced a theoretically bewildering choice.[30] Even if they were to not use any other source for watching films and series—an unlikely, conservative scenario given that VOD users tend to be among the heaviest consumers of all sources overall[31]—tens of thousands of hours of content would be available for

TABLE 1 AND FIGURE 2 Scripted Original Series

	Total	% Change Year on Year	Online	% Annual Change
2019	532	7	212*	33*
2018	495	2	160	37
2017	487	7	117	30
2016	455	6	90	96
2015	419	8	46	44
2014	389	11	32	33
2013	349	21	24	60
2012	288	8	15	150
2011	266	23	6	50
2010	216	3	4	300
2009	210		1	
(2002)	(182)		(0)	

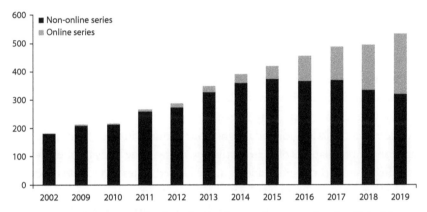

TABLE 1 AND FIGURE 2. Number of original scripted series produced in United States, 2009–2018 (with 2002 figures for comparison). Total commissions differentiated by series commissioned by online platforms and those by traditional television and cable networks. Asterisks denote that the proportion of online series for 2019 is estimated, based on Netflix spend and past increases. Percentage change values rounded to nearest integer. Sources: Daniel Holloway and Cynthia Littleton, "FX's John Landgraf Sounds Alarm About Potential Netflix 'Monopoly,' Overall Series Growth," *Variety*, 9 August 2016, http://variety.com /2016/tv/news/fxs-john-landgraf-netflixs-massive-programming-output-has-pushed-peak -tv-1201833825/; Lesley Goldberg, "FX Chief: Scripted Originals Set to Top 520 in 2018," *Hollywood Reporter*, 11 January 2018, https://www.hollywoodreporter.com/live-feed/fx -chief-scripted-originals-set-top-520-2018-1073348; Joe Otterson, "487 Scripted Series Aired in 2017, FX Chief John Landgraf Says," *Variety*, 5 January 2018, https://variety.com/2018 /tv/news/2017-scripted-tv-series-fx-john-landgraf-1202653856/; Lesley Goldberg, "Peak TV Update: Scripted Originals Hit Another High in 2018," *Hollywood Reporter*, 13 December 2018, https://www.hollywoodreporter.com/live-feed/peak-tv-update-scripted-originals-hit -high-2018-1169047; John Koblin, "Peak TV Hits a New Peak, with 532 Scripted Shows," *New York Times*, 9 January 2020, https://nyti.ms/2T8amRK.

selection. In late 2018, Netflix boasted twenty-five thousand hours of content in some territories, and Amazon nineteen thousand.[32]

The increases in the sheer numbers of films and series produced and distributed are not the only reasons for widespread perceptions of an audiovisual glut. The variety and number of dissemination channels have likewise counted as causes and symptoms of a fragmented marketplace: cable and satellite television and the oft-cited "500+ channel landscape." The steady offer of new VOD platforms (from industry giants Netflix, Amazon, Hulu, and iTunes to niche or regional offerings such as Fandor, MUBI, iFlix, or Einthusan) and their various business models (subscription, transactional, rental, ad-supported, and so on) make for an unmanageable and unmasterable experience. But even more traditional venues, like cinema screens and film festivals, have exploded.[33] One researcher has found nearly ten thousand film festivals that have run between 1998 and 2013 (approximately three thousand of which are active); 75 percent of these festivals began between 2003 and 2013. The United States and Canada hold the lion's share of these festivals, with an estimated twenty-one hundred festivals in North America. By all measures, these events have multiplied exponentially in the last twenty years, with estimates that there were fifteen to twenty film festivals in the United Kingdom in the late 1990s and roughly four hundred in 2017.[34] Extrapolating against the populations of these territories, there is one festival for every one hundred twenty thousand residents in the United States and Canada and one festival for every one hundred fifty thousand residents in the United Kingdom. These figures do not take into account the expanding numbers of "online film festivals," which often take place independently of physical events.[35]

Of course, one might speculate that the rise in films, series, festivals, and other channels of audiovisual product distribution simply reflects population rises or increases in the time that humans are spending in front of screens. The facts, however, do not substantiate such conjecture. The dramatic intensification of content production and distribution avenues must be read against only steady rises in population and mostly flat amounts of time spent per day on audiovisual media. The United Kingdom grew from 60.8 million to 65.6 million between 2006 and 2016, an increase of less than 8 percent.[36] Yet there was a 63 percent rise in the number of theatrical releases, even though cinemas enjoyed only a steady rate of 2.7 admissions per capita per year in that period.[37] The growth of series—each coming with tens, if not hundreds of hours of content—is even more stunning. Between 2002 and 2016, the US population grew by a little more than 14 percent, and cinema admissions did not keep up, falling from 4.4 to 3.8 per capita per year between 2007 to 2016. The number of original scripted series, however, ballooned by an astonishing 143 percent.

To be sure, there is clear evidence that VOD consumption is growing. Studies show that most people, especially the young, are consuming films and series in new ways: there is rising take up of free, ad-supported, on-demand content, less DVD and Blu-ray use, and relatively steady amounts of trips to the cinema. In turn, the amount of time spent watching linear television is gradually decreasing year on year.[38] Furthermore, consumers are now frequently subscribing to more than one SVOD service: in 2015 the average SVOD user in North America and Europe paid for 1.4 services, but by 2017 this number had ballooned to 2.6. By 2019, US households subscribed to 3.4 (and *used* 4.2) SVOD services on average.[39]

And yet, the most striking phenomenon is this: despite the explosion of content on offer, the average time spent on television, films, and streaming *in aggregate* and per capita is barely expanding; the overall amount of consumer spending on video entertainment is not rising significantly either.[40] Indeed, it would be difficult for most people to spend any more time watching recorded moving images without sacrificing sleep. One study found that on average Americans spend 55 percent of their time awake consuming entertainment media.[41] A 2018 survey found that UK residents spend on average 624 hours per year watching VOD—eight times as many hours as they exercise.[42]

So, while the number of scripted series has more than doubled since 2009, film production is spiking, the cable channel base has further swollen, and subscription numbers for VOD providers such as Netflix and Amazon have multiplied at staggering rates since 2012, humans are only spending a few extra minutes per day on these activities. It is easier than ever to produce media content and much more than ever is available in more venues; nevertheless, only a relatively constant amount is actually being seen. This leaves us with the sobering conclusion that even with the expansion of cinema screens, the proliferation of niche television channels, festivals, VOD platforms, and much-hyped cultural trends for box-set and SVOD-enabled bingeing, the sheer amount of content means that many films and series, despite often easy and instant availability, will not reach viewers. Among a surfeit of options, choice has become paramount.

This point is not lost on some perceptive cultural commentators. The "widening gap between limitless media and limited attention," according to media scholar James G. Webster's polemical formulation, "makes it a challenge for anything to attract an audience."[43] For Georg Franck, in a society in which content production has become cheap and time is scarce, attention is the "new currency," just as or even more important than money for media audiences faced with an inexpensive and often free flood of information.[44]

If there is more content than ever—and no corresponding expansion of aggregate media use—there exists an increasing social need to focalize attention. To be sure, part of the attention economy has led to identifiable aesthetic and cultural consequences, amplifications of long-term trends. Research has suggested that such production-led effects may include spikes in sexual, violent, or other provocative content. There is also evidence of an intense focus on narrative twists, tricks, or other gimmicks, especially in the first few minutes of programming, in order to maintain viewership in an environment of easy switching. Other studies have shown how editors are cutting films at ever-decreasing average shot lengths: audiovisual content is having to move at ever faster speeds to engage audiences. Correspondingly, on the distribution and marketing side of the industry, progressively sophisticated (and desperate) promotional practices have arisen: red-carpet antics and social media campaigns, transmedia storyworlds and found footage gimmicks, DIY crowdsourcing and promotional campaigns.[45]

Important means of focalizing attention in a flood of content extend well beyond content producers and digital marketers, however. Recommendation sources, and media users' own conscious and unconscious behaviors, remain equally, if not even more essential.

When faced with a plethora of media options and a finite amount of time, how do consumers choose? Webster—following Franck, Richard A. Lanham, and others—has developed a succinct and accessible account of the attention economy, in his idiom the "marketplace of attention," which increasingly determines the success or failure of media content and platforms.[46] With an endless number of outlets, more media content than any one person could ever hope to consume, and altogether easy (if not instant) access to these outlets and content, attention—eyeballs and ears—must be elicited and focused. Audience attention, a currency of limited supply subject to fierce competition, becomes the key to media survival and success in a "zero sum game that dooms most offerings to obscurity."[47]

Social scientists have made important contributions in explaining the parameters of media consumption decision-making. They dispel fallacies of rational choice-making agents, all-seeing, all-knowing super-consumers who are able to perfectly satisfy preexisting tastes and needs because they have perfect information and equal access to all media choices. Although digital media may "empower people in important ways, individuals are not the sole masters of their destiny. Media systems 'push' things at us, often in ways we scarcely notice. Contrary to most theories of choice, these encounters can cultivate preferences that would not otherwise exist."[48]

Some people assume that audiences seek out media content on the basis of *exogenous* preferences: in other words, fully shaped, more or less immutable tastes that have formed outside of and prior to our encounter with the media system. In this understanding, media are passive, disinterested objects, fruit waiting to be picked from a tree. However, in a world full of advertising and other more subtle ways to push content, such models seem naive. Many if not most preferences arise in at least partially *endogenous* ways: the media create interests and shape public attention.[49] Interests will ebb, flow, and change over time, stoked by information or recommendations about new films and series. Indeed, research demonstrates that the large majority of consumers are unable to clearly identify and articulate their preferences before beginning the media selection process.[50] Although even film and media scholars often speak in general terms about horror fans or sci-fi aficionados, empirical studies show that very few viewers exclusively consume one or two genres. Most audiences watch what they feel to be good exemplars of a range of genres, and systematically avoid a few disliked genres as well.[51] Cultural omnivores with tastes that defy strict diets, humans allow themselves to be pushed, to a certain extent, toward recommended products. (As we shall see, social effects of word of mouth and joint-viewing invitations are particularly strong "push" motivators.) In sum, "rational choice" of a consumer cognizant of all of his or her options is unrealistic in a digital media environment with limitless content. People maintain a "bounded rationality" when selecting films and series, a kind of decision-making circumscribed by the form of and access to media offerings, their own behavioral routines and biases, social herding, and other external guides that encourage, discourage, promote, or limit certain products.[52]

If media options are limitless and if media objects are experience goods, unique prototypes whose quality is difficult to judge before consumption, how can one ever hope to choose? There are essentially three ways. First, consumers adopt media repertoires, managing choice by restricting their focus on a much narrower range. Even though television viewers in many countries have access to hundreds of channels, studies show that in practice the vast majority will watch fewer than twenty in any given week.[53] The same is true of other delivery venues: you may discount the listings for all but the two or three cinemas closest to your home, you may refuse to subscribe to Netflix and premium cable channels, you used to skip the Classics section at Blockbuster and limited your focus to the New Releases aisle. Although there are millions of (film) news and information websites on offer, most people will only look at a handful. By limiting the number and scope of avenues for choice, audiences can help construct a personal media environment that seems more manageable, even at the cost of (and fear of) missing out on a particularly pleasurable film or series experience.[54]

The second means of narrowing choice is the use of heuristics, short-cut rules of thumb that allow individuals to rapidly categorize and evaluate whether the film or series requires further thought or can be eliminated as a possibility straightaway. Although economists remind us that films and series are experience goods and thus difficult to judge before consumption, in practice most people will not watch the full ninety minutes of a film or seven seasons of a series before deciding whether it is worth their time. Experienced channel surfers need less than a second to decipher whether the program will likely interest them. Distributors spend considerable money and effort on trailers, ads, posters, DVD covers, and VOD-thumbnail art because they know that the vast majority of potential consumers will size up their interest in the film or series purely by these means. Consumers are quick to categorize their available information, whether a two-second clip or promotional material, into assumptions about genre and production values, simplifying decision-making.[55] Despite the old saw, most of us do judge a film or series by the cover.

Finally, a third means of reducing choice is by referring to guides, searching for information, and consulting recommendations. Limited by the time, money, and barriers of access to go to festivals or upfronts, media users need to outsource part of their decisions to others. Consumers might read the year's ten-best-films list or consult the Emmy-nominated series; they can quickly gauge an unfamiliar film or series with a general internet search or a trip to Wikipedia or IMDb. A potential viewer might assess whether she wants to watch a film appearing on her Amazon page by remembering the lukewarm tip that her friend provided when he saw it in the cinema last summer. Or she might quickly check the review of her most frequently consulted newspaper. Of course, guides and recommender systems have their own meta-repertoires. Some users, particularly "maximizers," will read hundreds of reviews or search Google *and* Yahoo. Others may read one or two reviews, perhaps those in a trusted magazine, website, or blog or those whose notices arrive on the top of the search results page, or simply check out the Rotten Tomatoes or Metacritic score. A host of businesses—from the *New York Times*' Watching section to the original Jinni site or Watchly—exist as meta-recommenders that help consumers who have access to multiple SVOD services such as Netflix, Amazon, and Hulu focalize their choices.

Beyond looking to third-party recommenders on the internet, across social media, or through word of mouth, consumers go straight to the source. VOD platforms maintain their own built-in recommender systems, valuable means of ordering, classifying, and presenting media content beyond a top-ten or alphabetical listing. Explaining the main parameters of these increasingly important means of reducing choice in an age of media surplus remains the task of the next chapter.

TWO

How Algorithmic Recommender Systems Work

THE PRECEDING CHAPTER demonstrated how recommendations and thus recommender systems function as arbiters of cultural surplus and risk. This is the grand bargain of subscription-service VOD, which cannot exist without some type of recommender system: the two remain symbiotic entities and functionally indivisible. Linear television and cinemas drip-feed content; quantitative supplies of content are circumscribed by programmers but above all by the time and space of a broadcast or screening. Even the economics of traditional video shops, with their pay-per-time model and late fees, set limits. Subscription video on demand's all-you-can-watch smorgasbord would seem to kill the gatekeeper and unleash a flood of film and series. Yet recommendations exist as a check.

VOD suggestions—whether appearing as promises of curated selections (e.g., MUBI) or algorithmic calculations of personalized taste (Netflix or Amazon)—thus fulfill a social need: by presenting, among an insurmountably large possible selection of qualitatively differentiated experience goods, a manageable choice among the best available options. Of course, the concept of the proper selection varies among providers. Some promise an Arnoldian best of what has been shot and seen. Others shunt the user toward the already popular and proven. Still others appeal to the user's supposedly unique and discoverable taste. This spectrum of recommendation stances provides the subject for this chapter, which maps the formal, technological, business, discursive, and audience-experience coordinates of VOD recommender systems: the algorithmic and "human curation"–style positionings.

It is important to emphasize that the basic *social need* that recommender systems serve—having to choose among leisure pursuits—is a long-standing one. Considered in a non-techno-essentialist manner, antecedent methods, norms, conventions, and delivery mechanisms for suggesting films and series have existed ever since mass audiences could afford to consume them. Critics wrote reviews. Friends, passing acquaintances, and video store clerks told us whether a movie or show was "good," predicting whether it might correspond to our individual tastes. Posters and trailers and billboards and television spots beckoned us to cinemas or to our living room sofas at appointed times with stylish designs and attractive representations of familiar stars and fresh faces. The recommendation idioms had their own institutional norms, geographical locations or spatial conventions, temporal rhythms, and forms of intersubjectivity—conventions that anticipate and codetermine VOD recommender systems.

Naturally, just like Netflix or Amazon, these legacy recommendation forms also had (and still have) their own agendas and addressees. Posters and trailers emanated from the very agents who hoped to profit from the consumption of the product they advertised. The credibility of critics' reviews, in contrast, often derived from the textual and institutional posture of distance from industry interests, their "objectivity." Some recommendation outputs, moreover, were tailored to different markets: territories, geographical locations, language groups, censorship regimes, or identity-based demographic groupings such as age, gender, and even race. The best critics attempted to take account of the intended audience, looking beyond their own background and tastes to frame their reviews around the likely viewers of a superhero movie, a rom-com, an art film. At the very least they needed to keep their specific readership in mind: cinephile bible *Sight and Sound* is largely read for different reasons than the mass-market *Entertainment Weekly*, which in turn is often consumed by different people and for different purposes than a trade paper like *Variety*. At stake are shifting constellations of aesthetics, enjoyment, and economics, not to mention differences between broadcast and bespoke tips. These recommendation types persist, and their logistical and ethical implications still pertain; in an immediate-gratification on-demand world, however, some have been supplemented, bypassed, or short-circuited. Chapter 1 detailed the reasons that recommender systems exist among other heuristics and repertoires for choosing audiovisual content and why they have become increasingly necessary in the digital age. But, in general, the longevity of such forms (and their reincarnation in digital guise) bespeaks a basic truth: recommendations have a fundamental, constitutive, special, and unique role in the consumption of films and series. Recommender systems represent a new—but hardly unprecedented—way to solve an old problem.

To be sure, VOD recommender systems offer an apparent advantage over other means of reducing choice: the invisibility of how they push certain titles over others. In general, streaming services foreground their large catalog and the high quality of their selection in promotional rhetoric. This corresponds to a paradox of how people respond to cultural choice in general and to VOD recommendations in particular. When Netflix directly asked, audiences said that they desire above all "as much choice as possible" and "comprehensive search and navigation tools." Nevertheless, these same internal Netflix studies show that in practice users respond best to "a few compelling choices simply presented" rather than a massive library of titles listed A–Z or presented randomly in genre groupings.[1] One commentator calls this paradox the "user's dilemma": the more that users are "empowered" by easier access to more choices, the more they are forced to employ repertoires and heuristics to scroll and sort through the abundance—or rely on recommendations.[2] The very tools that allow SVOD platforms to sort, categorize, filter, and suggest content enable them to flood the market with content. Their service—to connect audiences to the best films and series among the plethora available— promises to solve a problem that the companies themselves contribute to creating.

The subscription VOD services that have had any degree of success, for all of their differences, have essentially followed this model: from Amazon and Netflix's data-driven front page (e.g., Netflix's forty rows of personalized genres) to MUBI's rotating catalog of thirty films. VOD recommender systems must project the illusion of plenty, but simultaneously prioritize or circumscribe choice by providing a set of recommendations in a credible, benign, and largely invisible way. Although audiences ultimately value the push of recommendations, they do not like to *feel* pushed—whether by ads or the appearance of limited selection. Indeed, as one scholar notes, audiences enjoy exercising their "free will" even within the confines of the recommendations, whether choosing among MUBI's thirty titles, or picking which of Netflix's fifteen recommended "Thrillers with a Strong Female Lead" to try out.[3] As we shall see, users' attitudes toward recommender systems depend on their beliefs about their own agency in relation to them, about whether they feel they are being "told what to see" or whether they can ultimately exert their own "free will" over suggested content.

The industry has roughly divided itself into two broad categories of SVOD services, which offer two modes of recommendation that are distinct in form if overlapping in purpose. On the one hand are the narrow platforms of smaller services, plentiful in number but miniscule in market share, which cater specialized content to niche audiences. Over the years these have included FilmStruck and MUBI (art cinema), Shudder (horror), Hopster (children's content, educational), Doc Club (documentaries), LebaraPlay (Tamil, Hindi, Turkish), Eros Now (Bollywood), or Afrostream (African). On the other hand are the usually much larger services that warehouse huge swaths of films and/or series, without a specific editorial focus and aiming at a generalist audience: for instance, Netflix, Amazon, Maxdome, iFlix. The former, narrowly branded services tend to use human-led curation policies to attract their target audiences. In contrast, the latter, which lack a coherent aesthetic or thematic brand, stress the breadth of their catalog, original or exclusive productions, and their algorithmic recommender systems in order to provide a personalized, satisfying experience for each individual subscriber.[4] Correspondingly, VOD recommendation exists on a spectrum: the extent to which users' personal data drive suggestions (and catalog acquisitions and even original productions). The source of this data includes user-generated ratings but especially consumers' information-seeking and viewing behaviors such as stops and starts, full and partial viewings, thumbnail mouse-overs, and compiling titles in watchlists.

Of course, every VOD provider uses data to some degree. Even MUBI and BFI Player, services that receive public subsidies, maintain cultural remits, and champion their "expert human curation" in promotional rhetoric, examine viewership information as a way to monitor usage and gauge future acquisitions. As such, we should be careful not to create too facile a binary.

Nevertheless, it is clear that some platforms use personal data more extensively in their recommendation services than others. Amazon, and especially market leader Netflix, use this fact as a prime selling proposition. They tout their recommender systems' ability to tailor to individual tastes by algorithmically processing viewers' data. These recommender systems promote content—by highlighting certain titles prominently and hiding others—predicted to more likely appeal to the individual user. These companies claim to use such data to acquire (or not acquire) certain types of content and even to use this information when commissioning original productions. In contrast, every user of MUBI receives the same presentation (with few exceptions, such as interface language) of the same thirty films. As such, these recommendation services offer the consumer a different experience, employ a different conception of taste, and rely on a different business model.

VOD recommender systems all seek to satisfy the same social desire: to focalize attention and reduce the unmanageability of cultural choice. And yet they come in a range of shapes that require different technological processes and suggest distinct ways of conceiving culture: broadcasting vs. narrowcasting; the best works vs. what works best (personalization); the wisdom of critics vs. the wisdom of crowds; the human touch vs. disinterested algorithmic precision. Elsewhere, I investigate the "human curation" side of the spectrum.[5] The present chapter begins the examination of the other end, exemplified by market leaders Netflix and Amazon Prime Video: algorithm-led, behavioral-data-informed methods of personalized recommendation. As a way to preview the case study of Netflix to come, I present here six of the most important characteristics of how such algorithmic systems work conceptually.

1. Data and Algorithms

Recommendations are produced by processing (viewers' personal) data through a series of algorithms. Despite their headline rhetoric of "human expertise," curation-style VOD services like MUBI or BFI Player collect viewership data and use algorithms in their platforms. Algorithms power their search functions; staff examine data on clicks and numbers of started and completed viewings to gauge the relative popularity of certain titles and make informed decisions about whether to program (for *all* users) films rather than series, French or Thai content, new releases or catalog classics, mainstream fillers or nonnarrative provocations. Nevertheless, this appropriation of data and algorithms pales in comparison to platforms such as Netflix and Amazon. The great claim of the latter is to manage complex computational processes that can predict what *individual* users might like, based on their vast compilations of data. These companies, unlike the curation-style providers, do not simply use algorithms as an added bonus, to determine in which language to display text or to send out a personalized reminder to finish a film started the previous week. For a Netflix or Amazon, the algorithmic digestion of viewers' data represents the recommender system's central task and main draw. "The provision of personalized recommendations," one computer science textbook on the subject instructs, "requires that the system know something about every user."[6] The purpose is twofold: to reduce perceptions of information overload and to foreground certain audiovisual content.

Personalized algorithmic recommender systems for cultural products began to appear in academic literature by the mid-1990s and were anticipated much earlier. They can be categorized into four basic types (content-based recommendation, knowledge-based recommendation, collaborative filtering recommendation, and hybrid recommendation) and the contours of their individual algorithms—after all, simply sets of instructions for computers to carry out—come in a (for non-experts) bewildering array of approaches to execute these commands: Bayesian networks, Markov chains, matrix factorization, latent Dirichlet allocation, and so on.[7] Although subsequent chapters will provide more detail on these approaches, the present humanistic discussion is less concerned with the minutiae of each algorithmic form. Rather, it is more pressing to illuminate how the computational processing of user-collected data seeks to solve the problem of choosing films and series in instrumental ways that both resound with, and differ from, legacy forms of cultural suggestion.

Algorithmic VOD recommender systems use statistical and machine-learning techniques to predict what content individual users will enjoy by one (or both) of two means: (1) detecting the similarities between the intrinsic qualities of individual films or series (called content-based or item-item recommendation) or (2) detecting the similarities between users' tastes (called user-user or neighbor-based recommendation, or simply collaborative filtering). Such systems learn and are thus supposed to become more accurate over the course of the item's shelf-life, the individual user's engagement, and the engagement of all users in aggregate. As any given user provides more data about what films or series he or she likes—either via explicit cues such as ratings or implicit clues such as which content he or she watches from beginning to end—the system will provide recommendations that are, in theory at least, better tailored to his or her taste. The system, in other words, quickly sizes you up and suggests content that, based on its data from other users, a person like you would enjoy.

The design of such systems begs legitimate questions about whether VOD services' amassed knowledge has any depth, about implicit biases, and about the translation of information into recommendations. A number of factors extrinsic to the algorithms themselves inflect the systems' perceived accuracy, such as the type of data the platform collects and how this information is interpreted. The meanings of behaviors, when taken out of context, can be misleading. For instance, does stopping content signal boredom or disgust? Or does it mean that you are especially interested and want to save the show for the weekend when you have more time to enjoy it? Streaming a film from beginning to end without pause could indicate that I am enthralled with the production. But it could just as well mean that I fell asleep or that the plodding score was a great way to get my baby to quiet down. A recommender system that relies on item-item connections (e.g., today's Netflix) must carefully weigh which qualities to emphasize and how to assign values to them. Will series be defined on their levels of romance or violence? What role, if any, will the setting or the protagonist's occupation or a cheerful resolution play? Will the production staff or the network on which the series initially premiered bear greater consideration or be ignored? The accuracy will also ultimately depend on the size and variety of the content acquisitions catalog and on a critical mass of behavioral cues, which in turn hinge on a sufficient number of human users. A small or monotonous selection or a platform with only a few dozen subscribers will produce imprecise recommendations, no matter how complex the algorithms may be. Does the viewer love Adam Sandler and need to be recommended his entire oeuvre at the top of the homescreen or was *The Wedding Singer* (1996) simply the least worst among the small pool of offerings available in the given geographical territory?

2. Personalization

The primary conceptual basis of the recommendation is its personalization *with respect to the individual user; the idea of bespoke and more accurate suggestions exists as a chief value-adding selling point. As such, these systems actualize foundational, persistent conventions of internet and new media talk.* All VOD services claim to offer appealing content. But whereas curation-style providers emphasize the quality of the *content* on offer, platforms on the algorithmic side of the spectrum place the individual *user* as the center point: recommendations come tailored to his or her taste. Considering the user's ratings or behavioral data, and deploying predictive models based on other consumers' preference patterns, the platform hides content of unlikely interest, pushing favorable content to within easy reach on the interface. In this way, algorithmic recommenders solve the surplus problem, focalizing attention not on the objectively "best" or even on the aggregately most popular, but rather on the user's subjective predilections. Just as VOD seeks to alleviate the dilemma of broadcasting (all viewers having to watch a single episode of program X each week at time Y), algorithmic recommender systems promise to resolve the dilemma of choosing experience goods. Diverse viewers maintain diverse tastes and come to consume content with different goals at different times: to laugh, to be moved or scared, to pass the time or learn something new. Recommender systems aim to beat the baseline of popularity and provide the personalized: predicting useful items for the individual, not the masses.

To be sure, recommendation modes have long taken audiences' previous viewing histories into account. It has long been a norm in reviewing practice, for example, to warn about spoilers, to make comparisons to previous films as a way to adjust expectations ("viewers who know *All the President's Men* [1976] will find *The Post* [2017] sentimental and lacking in dramatic tension"), to accommodate special constituencies ("only true horror fans can withstand *Jigsaw*'s [2017] gruesome onslaught"), and to cue a possible lack of sufficient background preparation ("don't attempt to see *Star Wars: The Rise of Skywalker* [2019] without first watching *Star Wars: The Last Jedi*" [2017]). Furthermore, personalized means of recommendation have existed for as long as there was cultural choice. Word of mouth, particularly as practiced by close friends and family, is an intimately individualized form of suggestion that bears the recommendee's viewing history and preferences (as known to the recommender or as elicited during the recommendation process) in mind.

Nevertheless, proponents of algorithmic recommender systems assert that the method is novel. VOD recommender systems, they maintain, provide a wider knowledge and superior level of detail. Unless you are personally acquainted with Eric Rentschler or Leonard Maltin or Chris Fujiwara—or *even if* you have a friend with encyclopedic knowledge of cinema and television history—it would be impossible to have immediate recall of thousands and thousands of films and series arranged not only by director or star, but also by level of romance or violence and so on. Indeed, unlike critics' reviews or advertisements—which have always taken into account audiences in terms of horror or rom-com fans and nationality, gender, or age—algorithmic recommender systems tend to consciously ignore such identity markers. As such, Netflix's and Amazon's VOD recommender systems represent the realization of the internet's early promises: content suited to the *individual* user's needs, rather than the broader public or traditional identity-based demographic groups such as race, sexual orientation, profession, or level of education. Some celebrate this development; others see this as an abrogation of culture, a remaking and recasting of individuals into new, algorithmically defined groups, incoherent and unknowable to users.[8]

Algorithmic recommender systems can tailor the information they provide in ways that short-circuit biases in viewers' choice heuristics and thus make the content seem more attractive to the individual. Research has long showed, for example, that festival prizes and awards, and in particular Oscar nominations, will make some people more likely to consume films; nevertheless, this information is ignored by—and can even have a turn-off effect for—other audiences.[9] VOD recommender systems can thus personalize different films' and series' cover art and descriptive information, alternatively (de-)emphasizing awards, critical reviews, various actors or directors, and story elements (e.g., romance vs. action) in order to entice viewers to watch content that—had it been cast in a different light—they would have avoided.

In theory, personalization helps consumers make more satisfying content choices. The strategy hopes to inspire user fidelity, increase sales, anticipate consumer desires, and better manage catalog acquisitions. VOD platforms that feature algorithmic recommendation engines deem such systems to be central to their competitive advantage, boasting of retained subscriptions and other rosy financial benefits attributable to personalized service. In turn, Wall Street investors value Netflix not as a distribution company or even in its new incarnation as a commissioning producer of films and series; rather, they see the outfit as a matchmaker between media content and users, akin to Uber's linking of taxi drivers and nearby passengers.[10] According to Netflix founder Reed Hastings, the key to Netflix is neither films nor series, but rather personalization: "A personalized service is the benefit of the internet. If you can otherwise do it offline, people won't pay for it online."[11]

Hasting's gambit is hardly unique to Netflix. Indeed, in order to understand the larger motivations and convictions behind the algorithmic recommender, one must understand the utopian new-media dreams and language of personalization. As media critic Michael Wolff writes, digital media "defaulted to the belief that information was the currency: after all, the new medium could provide information faster, cheaper, and with greater individual specificity; the functional dream from the early Web and then in essence put into mass practice by Facebook and Twitter was a newspaper just for you."[12] In the language of new media, fantasies of democracy, consumer empowerment, and cultural pluralism often intermingle with celebrations of bespoke personalization. This is the claimed fundamentally new—and wonderful—aspect of on-demand aggregators and their recommendation engines: the ability to access what you want when you want it, to binge or purge content unbound from the tyrannies of cinema listings and broadcast schedules. People would be freed from their analog shackles, digital evangelists predicted again and again. Innervated with new forms of agency and control over their lives and consumption choices, humans would be better able to fulfill their needs and live out their desires in custom-built ways.

The archetype for such utopian thinking about the internet's abilities to cater to personalized taste in the selection of cultural products remains Nicholas Negroponte's *Being Digital*. The MIT professor and *Wired* columnist's bestselling 1995 book deliberates on the future of digital technologies, business strategies, and consumer behaviors in the face of the coming flood of information enabled by digital data processing. Negroponte offers a convenient solution: "The answer lies in creating computers to filter, sort, prioritize, and manage multimedia on our behalf—computers that read newspapers and look at television for us, and act as editors when we ask them to do so."[13] One of the book's most striking passages envisions a computer "interface agent" that acts like a "well-trained English butler" to help consumers select among their many daily choices, from news stories and television programs to food and fashion.[14] As the algorithm learns the consumer's tastes, it can select content more and more tailored to his or her preferences, filtering out boring or upsetting news items, gauche outfits, uninteresting films, and irrelevant television shows. By employing an individualized filter, Negroponte suggests, consumers become their own newspaper editor-in-chief, television executive, and film programmer, heralding a more democratic society by bypassing authoritarian gatekeepers. Just as word-of-mouth recommendations are optimized when "expertise is indeed mixed with knowledge of you," Negroponte submits, interface agents "must learn and develop over time, like human friends and assistants."[15] For this reason, media must not attempt to narrowcast to a smaller subset, but rather curate for a unique audience of one: "Everything is made to order, and information is extremely personalized."[16]

Negroponte has been roundly and fairly criticized for his naivety about how special interests and large corporations would come to control and monetize these technologies, and how geo-blocking and other legal, regulatory, and acquisitions rights issues would foreclose his prediction that "information can become universally accessible."[17] Nevertheless, Negroponte's ideas about "personalized" aggregation, filtering, and curation serve as an enduring template, informing key aspects of the cultural discussion, business practices, and marketing rhetoric of today's film and series culture.[18]

3. Scale and Aggregation

Algorithmic recommender systems remediate traditional forms of cultural suggestion like word of mouth, but on a vast scale *and via the data-based* wisdom of crowds, *in other words, popularity among aggregated niche groups.* Algorithmic recommendation advocates maintain that these systems excel in the scope of both their inputs and outputs—and thus effectiveness and efficiency. The likes of Netflix and Amazon automate personalized word of mouth and project it onto a massive scale. By tagging all sorts of data (levels of romance and violence, settings, not to mention director, actors, and other information and criteria) about every title Netflix offers, the company has amassed a larger breadth of knowledge of film and series history—and its subscribers' aggregate use of that content, including where most users stop watching—than even the most learned critic or historian could amass in a lifetime.

Algorithmic recommender systems crave a distinct fuel: a critical mass of content and subscribers. As such, they rely on variables distinct from antecedent mechanisms and today's curation-style VOD services. After all, the recommendations of MUBI—a service that offers only thirty "heavily curated" films per month—would function more or less as well with one user as with one billion. Netflix and its ilk hinge not only on data and algorithms per se, but on the aggregation, diversity, decentralization, and independence of what has been called "collective intelligence" and the "wisdom of crowds": the explicit or implicit expressions of taste of all other system users. Providers and internet gurus imply that this base produces superior and more efficient forms of recommendations than those produced by one-to-one word of mouth or by expert curators and critics.[19]

Therefore, in a somewhat paradoxical way, personalization addresses the individual, but requires the masses: aggregations of data derived from the taste of crowds. "Collective intimacy," according to Joe Weinman, is a "new kind of intimacy, where a deep, value-adding individual relationship with each customer is based on insights derived from detailed knowledge of all customers."[20] This type of personalization remediates (and via computational processes can vastly transcend in scale) traditional forms of intimacy. Like hair stylists, the neighborhood butcher, or the bartender in the local pub, the systems aim to learn and intuit customers' personal tastes and preferences by observing behaviors across repeated interactions.

Machine-learning expert Pedro Domingos proposes three stages of recommendation intimacy, based on companies' size, scale, and technological prowess. In an initial mom-and-pop stage, a local business can know each customer personally and recommend products based on his or her needs. In a subsequent phase, as the company grows, it serves more people but not as effectively: the business can no longer respond individually to each customer and must use "coarse demographic categories" such as gender, age, income, or education level. A final stage of growth, however, allows the enterprise to expand and deploy machine-learning algorithmic technology to a critical mass of customers. At this stage, Domingos claims, companies such as Amazon, Facebook, or Walmart not only know individuals' tastes as well as in the mom-and-pop beginnings; they can also predict demand. "Learning algorithms are the matchmakers: they find producers and consumers for each other, cutting through the information overload."[21]

These observers deem algorithmic recommender systems to have fundamental advantages over other methods of personalization, benefits that yield lucrative pathways to profit. According to Weinman, algorithmic recommendation enables a new possibility of scale even as the required data collection (e.g., Netflix's tracking of starts and stops) often has zero marginal cost. "A customer-intimate individual such as a tailor, butcher, or doctor ... can only interact with and humanly remember a limited number of individuals, and there is an inverse relationship between quantity and quality." Such relationships—and the recommendations they yield—are *qualitatively* inferior because the limited sample available to these human recommenders hampers statistically significant insights. In contrast, algorithms can transcend such small samples: "scale is not only manageable but advantageous, because these subtle inferences are more statistically significant and thus more conclusive when based on more data points."[22] Algorithmic recommendations are *better* than your trusted butcher, bartender, or barber's tips because, this line of thinking claims, they are immune from human cognitive biases. They are not prone to fads or social herding and remain immune to common human tics in analyzing data, such as confirmation bias, illusory correlations, or the "hot-hand fallacy," which mistakenly interprets random events as related. "Computer algorithms," Weinman asserts, "can avoid all of these."[23]

Like a doctor with the capacity to treat millions of patients—and thus be able to detect larger pathological trends more globally and accurately—Netflix or Amazon can diagnose and predict patterns in cultural taste. Because the film and series catalog (changing ensembles of thousands of titles) and the user base (hundreds of millions of subscribers) are much too large for any one critic, programmer, or team to furnish with adequate suggestions, algorithmic recommender systems offer a superior method to matching content with users. Business historians see wisdom-of-crowds personalization, especially of the recommender system and the queue function, as the main factor in Netflix's rise to market leader, despite fierce competition from many upstarts as well as larger players (Blockbuster, the Hollywood studios). According to Willy Shih and Stephen Kaufman's Harvard Business School case study, rather than "making Netflix a difficult service to leave, Hastings wanted to make it a service that former customers would return to. Customers appreciated the personalized aspect of Netflix's service, a dimension that the company continued to improve. The proprietary recommendation system grew more accurate in predicting a user's taste as the number of films rated by a subscriber increased."[24]

In public appearances, press releases, and publicity, algorithm-led VOD platforms celebrate and promote the "collective intelligence" behind its recommendations. As one Netflix representative emphasized, "whatever you see on the Netflix service is actually a direct result of our members' preferences," recounting the thousands of hours of labor performed by a "about a dozen or two TV 'taggers' . . . employed around the world," who "tag all of our original content, second by second, over 1,000 tags that they can apply and that will get us a lot of information on what is actually in the programme and therefore we can actually make better recommendations compared to just whatever you view and the viewing habits of others."[25]

4. Nontransparency

Algorithmic VOD recommender systems depend on (personal viewing, user-generated ratings, and meta tagging) data collection and computational mechanisms that are integrated into the platform, often not transparent to the user, and treated as proprietary by the service. Because their basic function is to sort and suggest, to provide a means of access beyond "search," VOD recommender systems assume that the user's taste is at least partially endogenous: neither absolutely preexisting nor entirely unmalleable. The engines remain, by their objectives and activities, "pushers." They intervene into the market for attention by serving, but above all shaping, tastes:

delivering products that correspond to existing preferences, offering the popular, or providing new discoveries. They aggregate and filter, display and hide; by commission and omission, they program a palatable set of options ready for consumption. And yet many if not most VOD recommendation engines do not make these workings obvious or transparent in their consumer interface. Suggestions blend into the benign, one-stop-shop interface or remain partially veiled by the friendly rhetoric of discovery, quality, good works, or personalization. The secretive, proprietary business strategies surrounding the complex structures of the algorithmic recommender systems, their opacity to the general public, have generated concern among cultural commentators.[26]

Following Webster, media structures such as recommender systems and their larger technological apparatus, user interface, and institutional practices, can be relatively open or closed, obtrusive or unobtrusive. An open structure, in the view of some media theorists, is one that "allows people to consume, create, and share a varied diet of media," whereas a closed system has a design that filters out "discordant media encounters" and promotes cultural homogenization by shunting users into balkanized filter bubbles. Furthermore, an obtrusive media system maintains structures that are readily legible to the user, who is free to opt in or out of them. Unobtrusive structures, Webster writes, "work behind the scenes. Users may have no idea that they are operating within a structured environment," for example, if they are not aware that their search engine results, news feed, or VOD platform is customizing the respective displayed search results, articles, and videos with an algorithm.[27]

The following chapters examine the proprietary nature of Netflix's data collection and dissemination in great detail. But, generally speaking, algorithmic recommender systems tend to have relatively closed and relatively unobtrusive structural characteristics. In contradistinction to legacy recommendation forms as well as today's curation-style recommenders like MUBI or BFI Player, Netflix and Amazon closely guard—and, I will argue, monumentalize—the proprietary and nontransparent nature of the data behind the suggestion and the form and structure of the recommendation apparatus itself. Although one may suspect that any given film has been suggested because the financial terms for distribution were advantageous for the platform (just as Netflix and the other algorithm-based recommenders operate), in general a "curation"-based service such as MUBI or BFI Player makes plain the reasons for including the content on its catalog list or as a suggested title. The film synopses on their platforms stress aesthetic innovation, creativity, potential for enjoyment, or other positive terms that stress the *quality* of the production. As such, these descriptions seek to present a thorough case for why the film is presented to the user. They feature lengthy editorial context for recommendations—a retrospective on Douglas Sirk or a series on Chilean cinema, an anniversary or holiday ("romantic films for Valentine's Day"), or any number of written justifications that take on the form and tenor of capsule reviews or repertory cinema catalog entries. In contrast, algorithm-led recommender systems tend to omit the reasons behind the recommendations or truncate them into the barest, often explicitly personalized justifications: "Because You Watched *The Killing*" (2011–2014). Of course, there are some practical reasons for platforms such as Netflix or Amazon to omit this information: the technical difficulties of assembling lengthy and coherent reasons, the design and useability flaws of cluttering the interface with blocks of text. Personalization has its costs. But the consequence of omitting the whys and wherefores of recommendations serves to elide, by way of obtrusiveness, the very existence of the recommender system itself. In theory at least, an unassuming subscriber might believe that his or her homescreen features the same sample content, the same categories and rows, as that of every other member. (As we shall see in practice, however, many if not all users are at least selectively aware of personalization functions.)

The opaque and proprietary nature of algorithmic recommender systems is suggestive in several ways. It foreshadows the lengths to which VOD companies guard their systems as secret formulas of their success, mystify and promote them to consumers in publicity as their unique selling proposition (USP). Indeed, somewhat uniquely compared to historical antecedents, VOD recommender systems do not have a separate interface from their content delivery mechanism; points of access and suggestion merge seamlessly. Unlike a critic's review, advice from a friend, or even Rotten Tomatoes or Metacritic, recommender systems exist *within* the VOD platform; they are fully integrated into its user-facing built environment. These sites offer few if any visible boundaries between their exhibition and recommendation functions. From the user's perspective they are essentially indivisible, assuming that he or she registers the recommendations as such, that is, rather than a random or comprehensive presentation of content. Of course, we can identify precedents for merging recommendation and content delivery location: posters and trailers at the cinema, or ads, chyrons, and electronic program guides on television. But even these relationships between recommendations and exhibition offered more distinction: something you looked at or watched with a spatial (outside the cinema, on a foyer wall) and above all temporal (minutes or hours, but often days, weeks, or even months) remove from the act of consumption. Even if the distinctions between criticism and promotion, publicity and spontaneous peer-to-peer buzz, and ranking and curation were never as pronounced as asserted institutionally, traditionally a clear conceptual difference existed between advertising, on the one hand, and criticism and word of mouth, on the other. Whereas the former always flatters the product, word of mouth and especially reviews imply an independence from the economic imperatives of distribution. In this way, VOD recommender systems exist in a liminal space somewhere between advertising, criticism, and (aggregated) word of mouth. Because trusted critics' reviews and friends' word of mouth derive credibility from their distance from the content producer or their personal connection to the recommendee, respectively, algorithmic recommender systems must devise alternative means to engender trust.[28]

5. Credibility

Algorithmic recommender systems attempt to establish their source credibility *via promises of scientific objectivity, technological innovation, and design.* Ultimately, from the user's perspective, the most important feature of any recommendation type is not necessarily its level of mathematics or the precise algorithmic formulas, but rather the success of the suggestion and above all the credibility of the recommendation source. Studies demonstrate that users' perception of the source's credibility has a direct relationship with their motivation to follow the recommendation and the likelihood that they will share it with others.[29] Psychologists have shown how different sorts of suggestions will more or less persuade different recipients and yet, regardless of the medium of the message, credibility is paramount for the decision to follow the tip.[30] The chapters to follow will illuminate how Netflix and other providers attempt to establish trust via their publicity and means of selection and presentation, and how real users respond to these appeals. We will examine various instances and forms of credibility-building. The language of science, precision, and personalized wish-fulfilment is key for some and indeed for the most widely used engines. Other services, however, seek to exploit mistrust of big-data processing with appeals to the handcrafted, the bespoke, community, and friendly forms of recommendation expertise.

Of course, credibility has always been of central importance to legacy forms of cultural recommendation. A whole host of studies have intimated how different audiences will be more or less willing to trust individual recommendation sources (critics, television ads, print publicity, Twitter, word of mouth). Furthermore, within these categories people deem particular instances more or less credible. For example, word-of-mouth credibility depends largely on the message recipient's satisfaction with any prior recommendations but also in the perceived similarity between the recommender and him- or herself.[31] In other words: we are more likely to trust the advice of "people like us," however subjectively that concept may be understood. In turn, just as how one user may trust one friend's advice over another, individual consumers assess certain critics or magazines or websites as more or less credible. Critics have attempted to establish their authority via certain modes of audience address, claims of ethical distance from (or special insight into) the industry and its economic imperatives, stylistic performance (in print, radio, television, and other media), expressions of knowledge or expertise, and institutional association.[32] Grant Blank speaks of a "credibility continuum" between "connoisseurial reviews" (e.g., critics' film write-ups) and "procedural reviews" (e.g., *Consumer Reports* ratings of car tires, *Zagat*'s restaurant reviews, or box-office lists). The former's credibility relies on consistency, iterative trust-building (reading the same critic and comparing one's opinion over time), the appearance of impartiality, the reputation of the institutional context, the strength of evaluative arguments, and personality (expressed in rhetoric or, in the case of audiovisual reviews, appearance). In contrast, the latter's credibility rests on their mechanical, objective, and quantitative data-collection and testing procedures. Algorithmic recommender systems clearly seek to partake of the procedural review's credibility criteria.

The recently developed capabilities to target and narrowcast advertisements and other forms of recommendation has in general produced skepticism among consumers.[33] This epistemological conundrum—how people react when they know a recommendation emerging from their outgroup is nonetheless tailored to their previous personal consumption behavior—precipitates pressing questions for digital-age recommenders, whether Netflix's recommender systems, MUBI's curators, or the critics and lay users featured on Rotten Tomatoes. How does one create credibility, let alone cultural authority, among the vast array of information providers on offer? Where do, in other words, VOD recommender systems fit within the long-standing spectrum of information and suggestion sources? With whom might they enjoy more or less credibility and how do they influence users' viewing habits and tastes? The results of the empirical audience study, relayed later in this book, reveal insights that help answer these questions.

Clearly, the issue of recommendation credibility is crucial for the respective VOD businesses; they must engender this trust in users' successful engagement with satisfying films and series but also in their promotional rhetoric. And, as we shall see—despite widespread assumptions that users more or less blindly follow VOD recommender systems—empirical evidence suggests that a significant credibility gap remains. These systems attempt to accrue, but at best partially attain trust, which is usually limited to certain viewing situations among certain users, and even then, often with a healthy dose of ambivalence and skepticism. Most audiences say they do not (and do not *want* to) use and accept the recommendations of VOD services over other types of information and influence, such as word of mouth.[34]

Some of the most important criteria that audiences use to determine their trust in digital recommendation sources in general—and VOD recommender systems in particular—pertain more to the design, aesthetic experience, and useability of the interface than to technological forms or mathematical formulas. There is evidence to suggest, for example, that people find longer, moderate online (peer, WOM) film reviews more credible than those that are shorter or have extreme (whether very good or very bad) ratings.[35] Research indicates that few internet users laboriously research the backgrounds, principles, or any other aspects of websites or recommender systems. Instead, they use rules of thumb to quickly assess trust via the "predominant credibility consideration" of "'site presentation,' or the visual design elements of Web sites, rather than any content or source information."[36] Grammatical errors, typos, technical faults, pop-up ads (or other overtly commercial material), inconsistencies, and ugly color combinations are more detrimental to credibility than concerns about computer code or hidden political ideologies or economic imperatives. (We will see how real users talk about their preference for Netflix over Amazon above all because they perceive the former VOD platform to be superior in terms of design, layout, and useability—more so even than content factors like program selection or high-quality recommendations.) In addition, users deploy a number of other heuristics—for instance, relying on the reputation of the site, testimonials, or endorsements from friends or other (subjectively perceived) "like-minded individuals."[37] As a result of such indications, early developers of recommender systems for film, grappling with issues of credibility, attempted to simulate trusted-friend word of mouth as a way to encourage take-up of predictions.[38]

Today, recommender system designers use these principles when developing and evaluating the effectiveness of their programs; streaming services' CEOs seek to implement tricks of the trade in order to inculcate credibility and thus attract and retain subscribers. Computer science textbooks, citing the primacy of "the appeal of the overall visual design of a site, including layout, typography, font size and color schemes" in useability questionnaires, urge developers to attend to aesthetic features just as much as transparent explanations and recommendation accuracy.[39] Curation-style VOD services such as MUBI deploy sleek web design, modern textual style, and especially an idiosyncratic and us-versus-them conspiratorial audience address. These platforms also typically use longer and more discursive explanations (including references to awards and festivals and contextual historical knowledge) to justify selections and build credibility.

In contrast, algorithmic providers, although surely investing in the design and useability elements that consumers hold dear in their assessments of credibility, face another sort of dilemma. How does one enjoin feelings of credibility in collaborative filtering models—which link users with similar past viewing behaviors and tastes rather than any intrinsic content characteristics—that, by design cannot explain *why* a certain film or series is appropriate for the user?[40] In these cases, designers have sought to explain how the process itself mimics word of mouth or have subsumed such explanations under other rubrics (Trending Now; Popular). In practice, however, such justifications tend to confuse users rather than evoke trust: the "art of designing effective explanation methods for CF [collaborative filtering] is to provide the right abstraction level to customers," one textbook advises.[41] One recommender system prototype experiment tested twenty-one different methods of explanation. These included showing users a table of how "neighbors" (i.e., other users with similar tastes) rated certain films, simple statements about the past performance of the recommender engine (e.g., "MovieLens has provided accurate predictions for you 80% of the time in the past"), arguments related to types of content or actors or filmmakers, critics' ratings, or providing no explanation at all. Although the study found that offering too much information has negative effects on the likelihood of selecting the given content, it concluded that explanations, in general, and especially those that mimicked word of mouth (by providing neighbors' ratings) or referenced the effectiveness of the system in past performance, increased user confidence in recommendation credibility.[42]

Ultimately, algorithmic recommenders have sought to root the credibility of their recommendations in the technology itself, projecting a science-based promise of objectivity, a precision supposedly as universal as gravity or physics. Just as *TV Guide* must be correct each evening in order to be trusted—and even more so than a favorite critic who builds credibility over the years with insight or correspondence to the user's experience or taste—algorithmic recommender systems seek to convey a superhuman infallibility.

Proponents of these systems stress the disinterested objectivity of suggestions. Because of your reputation, self-identity, or, indeed, completely unbeknownst to you, you may give the impression that you cannot stand soppy melodramas, visceral action flicks, or steamy sex pics. In some cases, however, such beliefs may constitute virtue signaling or aspirational social climbing, rather than pure expressions of what you would prefer to spend the next two hours watching. The makers of Netflix's recommender system claim to know your deepest viewing secrets as they tabulate and view your clicks, mouse-overs, and false viewing starts to determine what *really* captures your attention. Indeed, they imply that these new automated systems may know your taste better than you (and your partner, parents, or best friends) ever could. Because so many derive data from behavioral cues rather than explicit ratings, the new recommender systems claim to amplify and purify word of mouth, freed from unconscious biases, pretentions, and other self-deluding public demonstrations of taste.

These systems' credibility, in other words, derives from the "science" of personalization, what media scholar Tarleton Gillespie calls "The Promise of Algorithmic Objectivity." "More than mere tools," Gillespie writes, "algorithms are also stabilizers of trust, practical and symbolic assurances that their evaluations are fair and accurate, and free from subjectivity, error, or attempted influence." He continues: "The *articulations* offered by the algorithm provider alongside its tool are meant . . . to bestow the tool with legitimacy that then carries to the information provided and, by proxy, the provider." The discourse of objectivity "certifies it as a reliable sociotechnical actor, lends its results relevance and credibility, and maintains the provider's apparent neutrality in the face of the millions of evaluations it makes. This articulation of the algorithm is just as crucial as its material design and its economic obligations."[43] Analyzing the performance of objectivity, scientific precision, and technological neutrality—across press releases, publicity stunts, trade discourse, tech magazines and blogs, published interviews, and in the interface itself—will be a major task of the Netflix case study in the two chapters to follow.

6. Taste

Personalized recommendations deriving from the data-based wisdom of crowds, rather than "editorial" content or traditional cultural gatekeeping by elite experts, might assume a fundamentally new form of cultural taste, a potentially altered purpose for the recommendation of audiovisual products. Whether masses of consumers' opinions truly constitute intelligence, wisdom, or even a credible source for a good recommendation is an issue of much contention. Some observers argue that "intelligence" is simply new-speak for "popularity" and that recommendations made on this basis are not disinterested assessments of cultural taste, but rather instruments that themselves help create popularity via social herding and self-fulfilling prophecies.[44] Regardless of one's conclusions on this matter, it is clear that the very processes of algorithmic recommendation not only operate within different parameters than their curation-style cousins; they deploy different *measures and goals* of successful suggestion, attempting to bypass traditional norms of cultural gatekeeping.

When I teach film criticism to university students, one of my first lessons is to jettison the notion of a "good movie." The question "Was the film good?" is perhaps the crucial one for most popular word-of-mouth engagement with films and series, but it elides too many crucial issues and conflates too many nuanced concerns: stylistic mastery, innovation, historical significance, authenticity, aesthetic achievement, or consumer fulfilment, to name a few examples. And yet the question at least implies an idea of culture and taste that transcends the demandee him- or herself. It envisions the given work as having potential beyond the individual, something sharable and applicable elsewhere, rather than a made-for-me prototype to be consumed and immediately discarded. The discourse of personalization—actualized in the rhetoric surrounding algorithmic recommender systems, the assertion that there are "100 million different versions of Netflix"—offers a distinctly different understanding of what film and series, indeed culture itself, should be and how taste functions in this constellation.

For most of history, for as long as we have records of humans reflecting on the value of literature or music or visual art, a certain idea about cultural taste obtained. Its essence was expressed in Matthew Arnold's pithy dictum of imparting "the best which has been thought and said in the world."[45] This formulation contains rich meanings, understandings, and assumptions. That there is an objective standard or at least relative consensus about what constitutes legitimate culture and good taste. That taste can elevate people into better lives and help society maintain peaceful order. And that a great work is equally great for everyone.

Arnold's ideas, as particular as they were to the intellectual and class undercurrents of the Victorian Age, represented the culmination of two millennia of thinking about cultural taste, from Plato and Aristotle to the Enlightenment-era grapplings of Herder, Hume, and Kant. For most of recorded humanity, taste—and the constituent sensual human functions charged with taking in information, experience, and pleasure—was considered above all as an essential part of judging and making sense of the world, a condition for sentience, a fundamental part of what made humans human and kept civilization civilized. To be sure, the subjectivity of taste was long a matter of debate, with some thinkers wedded to absolute values and categories of beauty, while others allowed for some degree of differentiation. And yet none seriously contemplated a snowflake storm of individualized relativity, of judgment produced by unique standards or even on whim.[46]

Of course, thinking men and women chipped away at this consensus long before the arrival of the internet. Pioneering sociological studies, by the likes of Thorstein Veblen and Pierre Bourdieu, suggested that taste has functions other than aesthetic fulfilment or even social cohesion: it could include conscious and unconscious self-classification, signaling meant to define the individual's in- and outgroup and to register his or her place in a pluralistic, hierarchical society of many subcultures. Some observers reflected on the constructedness of traditional artistic canons, noting that the heavy concentration of dead male Europeans and their white descendants on such lists was more than coincidental. Others wondered whether widely popular and putatively lower genres and forms—Westerns or detective novels or comic books or pop music—could not also contain aesthetic substance and social significance. Perhaps taste could not be prescribed. Perhaps it depended on historical contingencies and reflected sociologically comprehensible forces, rather than any absolute truth that could be discovered and then applied to all. On a micro-level, such studies suggested taste might not be a common good but rather subject to some individuation: one's social *subgroup*—or desired membership in a subgroup—strongly indicated whether one would prefer jazz, classical music, art films, or soap operas. And yet, on a macro-level, such conclusions were just as universalizing: membership in rigid categories like "lower-middle class" *determined* taste, not any one individual's unique and potentially mutable sensibilities.[47]

Indeed, even if the lofty language of great books and genius authors abated somewhat in a more cynical era where explicit deferral to authority became uncouth, the notion that there were things such as good reads or movies—works that somehow transcended individual predilections and could be recommended equally during a chance encounter on the bus as to one's trusted friend—persisted. Yet with the advent and development of the internet and other information-processing innovations, an explosion of easily accessible data, something more fundamental seemed to change. Applications meant to harness and establish order among the abundance of information brought with them new ways of seeing culture itself. Stated simply: commentators wonder whether we are shifting from broadcasted universal values toward a narrowcasted absolute relativity of taste, a brave new world in which human input into questions of evaluating culture may become inefficient and superfluous. To be sure, a whole host of industry insiders, tech columnists, and business gurus recognize and celebrate this step change in taste and wish-fulfilment as the triumph of democratic empowerment, the much-wished-for death of the gatekeeper. But many academics, in particular media scholars, see rather a sinister transition and irrevocable break, an abrogation of a sociocultural contract. The MUBIs and BFI Players of the curation-style VOD world clearly continue in the Arnoldian view of taste and culture. In contrast, the algorithmic recommender system à la Amazon and Netflix, many fear, represents a technological symptom of a transition in cultural taste from absolute truth to absolute relativism, a "customer is king" mentality that emblemizes less even the "wisdom of crowds" than the narcissistic pleasure of one.[48] This book will test these claims and the afterword will return to this discussion.

. . .

As a final distinction to curation-style suggestion styles, algorithmic recommender systems evoke by far the most concern in public discourse. In fact when commentators refer to "recommender systems," they inevitably refer to this type. This technology-first way of thinking neglects the fact that recommendation forms reside on a spectrum with more traditional forms of gatekeeping (MUBI, BFI Player, and so on), other ways of finding out about media (listings, reviews, ads, word of mouth), not to mention the plethora of repertoires and heuristics that individual viewers deploy to make decisions. Among the op-eds and comments, let alone the longer book-length squibs contra VOD recommender systems (and related phenomena, such as Facebook's newsfeed or Google's search engine), objections to (indeed mentions of) MUBI or BFI Player, film posters, newspaper critics, or neighbors' word of mouth are scarce indeed. This is true even though, as we shall see, the influence of these suggestions demonstrably surpasses the reach of VOD recommender systems. Because of this proliferation of commentary, I devote much of chapter 4 to this subject.

Now that we have explored the defining conceptual characteristics, social needs, and other imperatives behind the existence of recommender systems, we need to pursue a closer and deeper engagement with how they work in the messy practice of media markets, business organizations, and consumer behaviors. The following chapters will provide such a case study: world market-leading VOD service Netflix and its personalized algorithmic recommender system. But, by way of conclusion to chapters 1 and 2 and as an important guide to the discussion ahead, a few key points deserve reemphasis.

First, although VOD recommender systems depend on the rhetoric of novelty as part of their USP and deploy these discourses in their publicity efforts, their basic purpose fulfills a long-standing need for cultural suggestion and choice. As experience goods valued for unique traits and usually viewed only once, films and series have innate qualities that make judging them before consumption particularly difficult. Recommender systems are crucial managers of quality risk.

Second, although humans have long complained about information overload, in recent years there has been a demonstrably significant increase in the availability of cultural products like films and series in advanced industrialized countries such as the United States and the United Kingdom. In a media world in which the supply of options has long since outstripped humans' ability to consume, and increases in products are not met with concurrent increases in media time per capita, there exists an economy of attention.

Third, in order to reduce choice to a manageable size, cultural consumers engage in media repertoires and heuristics. Furthermore, a host of enterprising companies and business practices have emerged to harness and monetize this cultural overload, often using techniques of aggregation, filtering, and curation: in particular, recommender systems. These businesses that sell access to these devices promise to add value by directing customers to content that would have been inaccessible, inconvenient, unknowable, or otherwise lost in the surfeit of cultural products. Recommender systems are arbiters of surplus: the more choice VOD platforms offer and the more content they buy or commission, the more necessary recommender systems become for consumers to arrive at a satisfying choice.

Fourth, VOD recommender systems exist on a spectrum that spans two basic types: the best works (curation-style) vs. what works best (personalization); the wisdom of critics vs. the wisdom of crowds; broadcasting vs. narrowcasting; explanation vs. nontransparency; the human touch vs. disinterested algorithmic "objectivity." Personalized, algorithm-led recommender systems form a central part of the business models, branding, and promotional rhetoric of the market-leading VOD companies, especially Netflix and Amazon. They are also credibility machines, recycling the trust invested in legacy forms and laundering their opaque technology with the rhetoric of algorithmic precision, scalability, and disinterested objectivity. Before addressing user experience, it is vital to closely engage with how Netflix, the world-leading VOD platform, has designed a recommender system based on the promise of algorithmic personalization and scientific objectivity, a mode of credibility building that remediates several legacy recommendation sources.

THREE

Developing Netflix's Recommendation Algorithms

IN 2017, the *New York Times* issued a satirical illustrated essay about a woman named Beth coming to terms with what to watch on Netflix. The article's humor relies on consumer knowledge of various platform features such as menu categories ("Watch It Again"; "Continue Watching"), the percentage match feature ("98 Percent Match"), flaws in the personalization algorithms ("Beth must scroll past recommendations for 'Mean Girls 2' and 'Picture Perfect,' despite having zero interest in these movies"), the "Because You Watched" recommendation rhetoric, the profiles feature ("Beth is put to the test when she tries to find 'feel good' movies, but her ex Josh's favorite show ruins her algorithm"), and the platform's lack of selection: "Her first choice, 'The Devil Wears Prada,' is never available, and although she has the money, she won't rent it on Amazon."[1] The subtext to the piece—beyond the social commentary that sees a young woman's life through Netflix recommendation idioms—is that the company, the world's most-subscribed video on demand platform, has fundamentally reshaped how Americans select films and series.

Indeed, there is evidence to suggest that Netflix's dominant market position has catalyzed aesthetic shifts (more series, documentaries, thirty-minute dramas, interactive content), uprooted industry norms (bypassing theatrical windows, dominating production schedules and award shows), and altered social protocols in consuming films and series, notably and succinctly described in the resonant anthropological scenarios "Netflix and chill" and "binge viewing."[2] The latter is an activity to which 79 percent of the UK public admits, incentivized by entire series premiering online in one-off dumps.[3] Billing itself as "the world's leading internet entertainment service"[4]—and as of 2021 boasting over two hundred million subscribers and more US customers than the country's largest cable provider[5]—today's

Netflix presents itself on its website, in interviews, and in employees' public appearances with a pitch reduced to three core messages: accessibility ("Available Everywhere"), on-demand rather than linear broadcasting according to schedule ("Consumer Control"), and the recommender system ("Personally Relevant").[6]

According to *Washington Post* television critic Hank Stuever, such messages veil the lack of an overarching and coherent sensibility, aesthetic, or identity; Netflix has neither the defining values nor the branding of an HBO ("It's not TV. It's HBO.") or even an FX. Indeed, instead of a vision, according to Stuever, Netflix proposes merely a race to deal, churning out a steady flow of press releases proclaiming subscriber numbers, new series starts, and poached executives from more established organizations. Its self-identity is that of a rogue thief or mischievous disruptor, a publicly traded revolution: transforming television and social norms, beating Blockbuster and then the cable networks and cable itself. Trigger-happy to renew middling series and favoring quantity over quality, the critic writes, the company's USP is that "more is more."[7]

And yet, *pace* Stuever, Netflix's calling card as a business is neither an aesthetic nor a genre nor even a sensibility: the company's chief selling point is its something-for-everyone-style recommender system. In a revealing 2012 interview, chief content officer Ted Sarandos argued that Netflix original production intentionally lacked a unifying style or form; indeed, cultivating an easily recognizable look or genre foci would contradict its intended center-ground, broad-church appeal. Netflix's trademark is neither horror nor sci-fi, nor is it even a more abstract characteristic such as quality. Its brand is personalization:

> I don't want our brand to influence our programs, and I don't want the programs to influence our brand. Netflix is about personalization. Making our brand about one [type of programming] over another risks polarizing our customers. Tastes are just way too broad for us to even consider it. If you ask five people what they love about Netflix, they will give you five dramatically different answers. So we have to be really careful to ensure our brand is really about the shows *you* love, not about the shows we tell you about.[8]

This message is stressed again and again, in ways both subtle and startlingly direct. Returning to Netflix's website, the main pitch pertains to its recommender system and how the data it collects helps the platform deliver content. "Freed from these constraints, Netflix is able to program for over

93 million unique tastes of entertainment lovers around the world, from the romantic comedy lover in Paris to the sci-fi aficionado in São Paolo and everyone in between." Boasting in 2017 of delivering "the biggest and most diverse slate" of content in the company's history, including one thousand hours of series, films, and specials, Netflix marketing rhetoric clearly focuses on accommodating individual tastes and curating personalized viewing experiences. Eschewing "one-size-fits-all" programming and the old media business of filling scheduling grids, Sarandos announces the goal: "We're catering to a dynamic world full of people with different tastes – and as a result, no two Netflix experiences are alike – we want to have something for everyone to enjoy." In the press release, vice president of product Todd Yellin chimes in on message: "Human beings have incredibly diverse and unique tastes—each person is more than the demographic group they belong to. At Netflix, we not only have a catalogue that meets the needs of these tastes and moods, but we use our technology to ensure we surface the right story to the right person at the right moment." What is that technology, in the formulation of the company? Mysterious, powerful, yet benign "algorithms" and a "global recommendation system," which delivers "an even more personalized experience that helps members easily find a diverse set of movies, TV shows and more that they otherwise may not have been able to discover."[9]

As we shall see, this rhetoric constitutes one part of a grand narrative, an unholy pact between Netflix executives, novelty-hungry tech journalists, and shareholders. A deeply rooted mythology, it focuses on surging subscription numbers and stock prices, ever more Emmy Awards and Oscar nominations, and skyrocketing spending on original content. A discourse of tumescence that inevitably partakes of the semantics of revolution and conquest, the framing seems to supersede logically the news media's traditional fixation on theatrical box office.[10] Film critics, media columnists, and business and technology reporters alike cling to metaphors about Netflix "declaring war on cinema" or otherwise imply that the platform is winning some zero-sum battle against movie studios, television networks, cable companies, and cinema chains, unquestioningly ingesting and regurgitating the company's own bilious self-image of an Establishment sniper.[11] In short, all sides seem to stipulate that Netflix and its recommender system represent fundamentally new and powerful phenomena, an unprecedented mode of cultural production, distribution, and suggestion.

This chapter scrutinizes the development of the Netflix recommender system (sometimes called "Cinematch," especially in early internal discussions), which purports to present aggregations of films and series tailored to individual users: personalization carried out under the banner of big data. In the way that the company challenges traditional modes of cultural mediation, taste formation, and information acquisition about films and series, Netflix's most important functions may be less film and series distribution, exhibition, or even production. In crucial ways, the platform operates as a vehicle of information and evaluation. Expressed visually via menu presentation, descriptive language and images, and suggestion emails—and working in tandem with other features such as the queue/watchlist and the search function—the recommender system is a tool of surplus management, a focalizer of attention, and so much more. Central to the company's cultural resonance and economic survival, the recommender system has been and remains Netflix's self-professed "key pillar" and "most valuable asset," with a ten-figure valuation.[12] Despite the focus on revolutionary or at least novel technological practices, however, the following case study demonstrates how Netflix's recommendation styles and designs derive largely from past forms, a promiscuous concatenation and remediation of legacy recommendation sources.

This first half of the chapter begins by explaining the origins and historical development of the recommender system. In competition with then-rival Blockbuster, the earliest design of the recommendation engine used video-store classification logic as a model and attempted to simulate video-clerk word of mouth. By the end of this early phase, Netflix had abandoned initial attempts to emulate broadcast criticism and traditional film-world categories (genre, stars, directors) and developed a collaborative filtering model based on automating peer-to-peer word of mouth.

The second half of the chapter proceeds to examine the redesign of the system for the introduction of streaming, above all the new capacity to collect data by surveilling behavior rather than relying on users' self-assessments of taste. This new design sought to build credibility with enhanced explanation techniques. Ultimately, I will argue, despite Netflix's persistent consumer-facing rhetorical focus on personalization as its brand and key feature, company engineers have optimized the recommendation engine to blend personalization with factors such as critical acclaim, short- and long-term popularity, novelty, and diversity, consciously copying and transposing attributes of newspaper criticism, top-ten lists, and word of mouth. In order to engage viewers' attention and engender credibility, Netflix's most successful personalization tools remain resolutely unpersonalized.

The Netflix story is a narrative of monetizing, automating, and personalizing legacy forms of cultural suggestion. Reminiscing in 2019 about the company's origins, cofounder Marc Randolph recalled his initial ideas for a personalized shampoo or dog-food business; films and series were an afterthought to other first principles: "I really only had a single filter [for the start-up]: I wanted it to be selling something on the internet and I wanted it to involve personalization."[13] In Netflix's own corporate history, told in a bullet-point timeline on its website, the introduction of the recommender system is highlighted as one of the fifteen major events since its 1997 founding: "2000—Netflix introduces a personalized movie recommendation system, which uses Netflix members' ratings to accurately predict choices for all Netflix members."[14] Despite the many changes to Netflix's mode of content delivery, offerings, and price point, as well as dramatic shifts in focus and format, recommendation has been persistently integral to the company since its first hours.

As the company has transitioned from DVD to streaming (outside the United States, all Netflix accounts are streaming-only, and as of 2019 a mere 2.7 million DVD subscribers remained), the recommender system has only gained importance.[15] Internal discussions cast personalized recommendation as one of the company's "keys to success" and as a primary means to differentiate itself from competitors. Netflix claims that 75–80 percent of streaming hours derive from its suggestions; since customers—according to executive Todd Yellin—will at most consider forty to fifty titles before selecting a film or series to watch, it remains essential to provide viewers with an attractive sample of that size at all times. Indeed, business pundits ascribe Netflix's customer retention primarily to its "insightful" recommendations, which they claim also reduce acquisition costs, decrease the need for advertising and loss-making free trials, and improve sign-up logistics. As previewed in the preceding chapter, collective intimacy and personalization feature among the most important value-adding processes in the contemporary distribution of audiovisual products and indeed the digital economy writ large.[16]

It is important to note that Netflix's recommender system remains an organic, historically changing set of operations, hundreds of algorithms that have been tested and tweaked over time. The chapter will discuss some of its technical properties (that is, those that have been publicly revealed: many are not). Nevertheless the aim here is not to exhaustively rehearse each algorithmic specification, but rather to chronicle the recommender system's major changes over time and analyze what this narrative reveals about Netflix's politics of information and suggestion, its conception of taste and user agency, its business model and marketing strategies. In this undertaking we must be mindful of the methodological admonishments of Lisa Gitelman and resist "the idea of an intrinsic technological logic": ready-made, pre-cooked conclusions based only on form.[17] Nuanced analysis requires thorough consideration of the technology's economic motivations, social settings, and cultural effects. We must endeavor to understand what instrumental uses the recommender system offers and what needs it seeks to satisfy, note its innovations, name its antecedents, and trace its genealogy.

According to Gina Keating's inside-access account of the company, in the late 1990s the Netflix founders enumerated a few value propositions that would serve them well against what they considered to be their main competitors, Amazon and Blockbuster. After all, Amazon was already becoming a dominant player in the online retail world and counted among its strengths a talented back-end staff, strong postal logistics, and an effective web-based transactional model. For its part, Blockbuster, as the largest American video rental chain, maintained massive economies of scale, preferential pricing deals with the Hollywood majors, and rapid (if not instant) gratification. Unlike under the Netflix model, customers could collect Blockbuster rentals from one of thousands of shops; the chain boasted that 70 percent of Americans lived within ten minutes of a local branch.[18]

The Netflix recommender system's development evinces attempts to transpose legacy recommendation structures and idioms, so as to offer a familiar yet superior customer experience, and challenge market rivals. The platform's original sales pitch revolved around content variety (to "boast the largest selection of movies on DVD in the world"), aesthetic design ("the customer interface had to meld the familiar layout of video rental store with the pictorial and descriptive come-ons of a catalog to make the merchandise seem worth the wait"), useability (the "ordering process had to be easy," indeed, less complicated than fetching a film from a physical store and returning it) and, of course, personalization (the site needed "to be a personal experience, as if each customer opened the door to find an online video store created just for him or her").[19] Of these features, personalization was key: taking advantage of the fledgling platform's online strengths, it served to clearly distinguish the upstart from Blockbuster's blandly anonymous stores. Netflix's search engine, called FlixFinder, enabled users to find films by title, actor, or director;

FilmFacts linked to important contextual information such as synopses, ratings, cast and crew lists, and DVD characteristics. Another application, Browse the Aisles, allowed customers to sift through lists of films classified by a genre or theme. In a feature that previewed today's recommender system, entering a favorite title yielded suggestions of related content. All of these informational functions seamlessly took over traditional recommendation roles, such as consultations with the video store clerk, word of mouth from friends, and looking up information from one of the many popular reference works of the era, above all critic Leonard Maltin's annual *Movie Guide*. The Browse the Aisles feature alluded unabashedly to the way in which navigating the site's genres replicated the architectural-classification model of video stores like Blockbuster.[20]

Keating reports that Netflix's early development team led by Mitch Lowe (who would later found Redbox) sought to create a "digital shopping assistant." Complete with a photo, name, and personality, the assistant would guide users' film selections by providing personalized advice that would correspond to the user's individual tastes. Although Lowe supposedly summoned the idea from his background owning a ten-store video rental chain—where browsers preferred the younger clerks' selection advice over his—this sort of anthropomorphic recommender clearly recalls Negroponte's digital butler.[21] That Netflix entertained the idea of a digital shopping assistant demonstrates how it sought to simulate word-of-mouth tips: a means of suggestion familiar to most if not all film and series viewers. The company's ultimate abandonment of this idea, however, indicates executives' desire to implement a system that would be less obtrusive (in Webster's sense), arguably less transparent, and one that might seem more impartial or scientific. Netflix later succeeded in implementing these qualities in subsequent iterations of the recommender system and emphasized them in its publicity.

The Netflix website launched in April 1998 without a comprehensive recommender system that could predict user tastes based on the collective intelligence of collaborative filtering. Nevertheless, company developers purposefully designed the initial platform so as to accommodate such a system once the required resources became available. Already in 1997, the founders envisioned a mechanism that could suggest films based on previous rentals and prior searches.[22] Sarandos, reminiscing in 2012, deemed this foresight to be Netflix's ultimate competitive advantage: "During the early days of the internet, when everybody else was spending big money on Super Bowl ads, we were investing instead in technology, on taste-based algorithms, to make sure every single user had a personalized, highly effective matching tool to use when they visited their site. For us, that's why breadth matters. We are trying to match tastes, and tastes are really specific." Sarandos continued: "Our website, which is so personalized, will help you find something that you're going to love. . . . We are restoring a sense of connection between consumers and content. I think audiences have lost that emotional investment in content because television can no longer provide them access in the way that they want it, or in a way that matches current lifestyles."[23]

From the start, Netflix's recommender system served an existential goal: to manage inventory within a delivery system in which content was finite. In contrast to the later streaming model, the company could at this point only offer customers the limited number of DVDs of any one particular title: each copy had to be purchased and stored in warehouses. This problem was especially acute for new releases, whose demand was especially high. Indeed, beyond personalization, the recommender system helped solve another serious issue: diffusing customer dissatisfaction with lengthy waits for unavailable titles. In the early years, Netflix faced considerable challenges in acquiring stock. Unlike Blockbuster, which had negotiated cut-price deals with the Hollywood majors for popular new releases on account of sheer buying volumes, Netflix found that studios would not lower the $15 wholesale per-DVD price. For this reason, the company could not acquire the copious copies of each new release necessary to satisfy consumer demand. In addition, new releases enjoyed significant external publicity in terms of advertisements and critical reviews, stoking consumer awareness and further skewing demand toward them. The company decided to de-emphasize new releases by nudging users toward older and less popular offerings, for which it would not have to invest as much to build up short-term inventory.[24]

Netflix used a variety of presentational and informational tactics in order to drive demand away from blockbusters. The search function and, later, the recommendation engine were programmed to show only those titles that were actually available to be shipped, rather than disappointing users with the tease of out-of-stock films. Marketing research revealed that many early adopters (the service first rolled out in the Bay Area) were Asian students with specific tastes, and the company bulked up on Bollywood, Japanese anime, and Chinese kung fu films as a way to service local predilections. Netflix began a series of promotional campaigns that focused on seasonal or time-sensitive content, such as films about upcoming holidays, anniversaries, or current events; other website features highlighted certain actors or directors and above all the films for which the company had the best financial terms and most available copies. An editorial team began writing reviews of films-of-the-week, obscure or forgotten titles that could unclog distribution bottlenecks by rebalancing demand.[25]

These modes of recommendation were only partly effective, however. Such promotions successfully shifted some attention away from new releases and toward less popular titles. But because the messages were broadcast to all users—rather than targeted to selected, particularly interested individuals—they simply created new stampedes around these alternatives and produced the same sorts of dissatisfaction. In response, Netflix attempted to dilute the concentration of interest by increasing recommendations to five per day; the increased output helped little, however. "Eventually," according to erstwhile chief product officer Neil Hunt, "we realized that the promotional value of writing the editorial blurbs was zero."[26] Given Netflix's means of distribution, the video clerk/critical review recommendation paradigm proved unworkable, an insight at which Amazon's retail unit was also arriving at that time.[27] Personalized messaging was necessary to solve the inventory issues within a delivery system that shipped physical copies via post; users could not all receive the same promotions simultaneously. In order to pique interest in older and less popular films and thus spread demand more evenly across Netflix's supply, individualized suggestions were essential.

Development of the automated, algorithmic recommender system, which they called Cinematch, came with several trials, corrections, and false starts during copious market research and A/B testing that Netflix undertook on its users. The company initiated the personalization process by having subscribers fill out a survey to identify favorite genres. Perhaps most importantly, it introduced a user-based ratings model: members evaluated content on a scale of one to five stars. The system was designed so that the more films the user rated, the more accurate the predicted recommendations would become. In addition, because it filtered results based on inventory demand, the recommended films were more likely to arrive promptly and thus buttress user satisfaction. Netflix found that once the system was fully implemented and refined, new releases made up only 30 percent of total rentals in 2006; by comparison, new releases comprised 70 percent of Blockbuster's rentals. The recommendation engine, according to business commentators, was the "game changer" in Netflix's survival and Blockbuster's demise. It was a reason to subscribe to the upstart service, but also a reason not to leave; users' switching costs included losing their queued titles and ratings, the basis of suggestions. Internally, engineers tested its effectiveness and found that users rated individually recommended films three-fourths of a star higher than new releases. Netflix's personalization tactics seemed to triumph over merely promoting the popular.[28]

In these early days, Cinematch contained serious flaws. These developmental issues strikingly reveal how engineers made initial (false) assumptions about cultural preferences and how they experimented with remediating legacy forms of suggestion. Their first idea yoked user tastes to elements intrinsic to the films themselves. That is, if a user awarded a film featuring Meg Ryan or made by Steven Spielberg four or five stars, the interface would respond by presenting several films in which Ryan appeared or that Spielberg directed. Programmers borrowed this idea from the consensus classificatory logic of legacy forms such as experts' books and video-store aisle signs, which were generally sorted by time period, broad genre (horror, sci-fi, action, rom-com), star, or auteur. The development team rapidly realized, however, that actors, directors, and similar item-based variables poorly predicted taste; users often loved one film and hated another produced by the same or similar cast and crew. (Netflix was confronting film's "nobody knows" experience-good problem in miniature.) Cinematch designers thus decided to forgo reverse-engineering taste via human intuition and submitted to a computational logic with essentially invisible criteria.

The gambit behind the revised recommendation engine remodeled the suggestion source and recipient: from the paradigm of a single critic broadcasting to all users—which had caused stampedes and imbalances in inventory—to a system in which all users' tastes inflected recommendations narrowcast to each individual. "We had this realization that if we gathered together a really large group of people, like thousands or millions, they could help one another find things, because you can find patterns in what they like," one early developer submitted. "It's not necessarily the one, single smart critic that is going to find something for you, like, 'Go see this movie, go listen to this band!'"[29]

Rather than intrinsic characteristics like genre or star, this new collaborative-filtering procedure prioritized linking *subscribers* who rated films in similar ways: they formed a "mentor group."[30] Subsequent films highly rated by one member of the group but unrated by others could then be recommended to the others, with the assumption that the content had a strong probability of appealing to other group members. The computer scientists who developed collaborative filtering consciously attempted to mimic and automate peer-to-peer word of mouth, one of the most trusted and effective forms of cultural recommendation.[31] In other curious ways, the idea of taste proposed in Netflix's "mentor groups" and in user-user collaborative filtering systems resembles insights Pierre Bourdieu proposed several decades earlier. The sociologist famously contended that—despite individuals' protests that their tastes were unique—group membership, especially class background and aspiration, strongly shaped, if not determined, cultural preferences.[32] After all, both Bourdieu's study and Netflix's initial user-user collaborative-filtering system assume that taste can be articulated to a larger group after deciphering any individual user's baseline of taste. Similar people, in other words, rate films similarly. In this model, taste is constant, stable, and emanating from the individual, even if related to a larger group. The significant difference to Bourdieu's studies of taste, however, is that the system's determination of "similarity" relies on a small sample of ratings and that taste clusters may have little or no bearing in terms of traditional demographic classifiers such as age, gender, race, ethnicity, place of residence, nationality, or indeed class. Collaborative filtering's personalization is thus, according to one commentator,

> less personalized than a store clerk. The clerk, in theory anyway, knows a lot about you, like your age and profession and what sort of things you enjoy; she can even read your current mood. (Are you feeling lousy? Maybe it's not the day for 'Apocalypse Now.') A collaborative-filtering program, in contrast, knows very little about you—only what you've bought at a Web site and whether you rated it highly or not. But the computer has numbers on its side. It may know only a little bit about you, but it also knows a little bit about a huge number of other people. This lets it detect patterns we often cannot see on our own.[33]

Netflix's new collaborative-filtering model had its own drawbacks, however. To be sure, such user-user systems are efficient, technologically stable, and relatively simple to implement. The source of their authority—other people like you—makes recommendations easy to justify and therefore potentially more credible. Nevertheless, from a logistical standpoint, they are more difficult to scale and remain sensitive to sparseness and "cold starts." That is, users who rarely rate films will tend to receive less accurate recommendations; content that is rarely rated will seldom be recommended and could be suggested to the wrong users.[34] Risks abounded, above and beyond presenting users with films that failed to captivate their attention and satisfy their tastes. Because pure collaborative filtering systems are agnostic about content, let alone cultural sensitivities, poor programming can produce potentially offensive results. Then competitor Walmart.com apologized and abandoned its recommender system after a Black History Month cross-promotion saw consumers of Martin Luther King documentaries receive advice to buy *Planet of the Apes* (2001) based on their prior purchases and tastes.[35]

In order to achieve what would be later called "the world's best collaborative filtering system," however, the company recognized the limitations of these early methods and the further innovation required.[36] In an effort to improve the precision of the recommender system—and simultaneously create brand awareness and trust—the company ushered in a science competition, the Netflix Prize (2006–2009), which awarded $1 million to any outside team of developers who could improve Cinematch's accuracy by 10 percent. Another contemporary endeavor, however, would prove to be much more crucial: the transition from posting DVDs to streaming video, which made drastic changes and improvements to the recommendation engine both possible and necessary.

"EVERYTHING IS A RECOMMENDATION": THE DEVELOPMENT OF THE POST-2012, POST-STAR CINEMATCH

"Streaming," according to Netflix developers Xavier Amatriain and Justin Basilico, "has not only changed the way our members interact with the service, but also the type of data available to use in our algorithms. For DVDs our goal is to help people fill their queue with titles to receive in the mail over the coming days and weeks; selection is distant in time from viewing, people select carefully because exchanging a DVD for another takes more than a day, and we get no feedback during viewing. For streaming members are looking for something great to watch right now."[37] The new consumer offer was compelling: an all-you-can-watch subscription model that bundled one hundred thousand hours of legally distributed audiovisual content, conveniently accessible to start, stop, and resume at any time, on nearly every device.[38]

These public statements articulated a fundamental truth: the transition from DVD-by-post to streaming did not merely constitute a new logistical delivery system or technology. It invited new social protocols in terms of how users made consumption decisions and indeed what sorts of content they would prefer. Trialing or sampling videos before settling on a choice—an extremely costly behavior if one needs to send the DVDs back and wait days before new titles arrived in the mail—was heavily incentivized under the new regime. Each individual title could be tested with essentially zero marginal cost, content was immediately available, and the viewer alone determined the film or series start (and pause or stop) time. The prospect of watching four, eight, twelve episodes of a series in a single setting—improbable if not impossible when limited to three physical disks at a time—suddenly became very real and indeed de rigueur.

The service's basic temporality transformed from a wish-list model of a ranked queue, stretched out over a period of days, weeks, and months, to an immediate-gratification machine with unlimited sampling. In the words of Neta Alexander, this represents a shift from privileging the "future self"— that is, the mode of self-understanding that people used when placing films on their queues with the knowledge that they would be consumed weeks if not months in the future—to the "present self," an identity based around instantly satisfying the whims of the moment.[39]

Streaming and its associated social protocols allowed for fresh data sources to gauge viewing behaviors, information that could fuel an expanded and more responsive recommender system. In addition, however, it called for different sorts of content. "(Mostly) gone are the days that customers would fill their DVD lists with artsy indie films or all of the Academy Award-winning documentaries they could, only for them to remain in queue purgatory," one commentator presciently foresaw. "Netflix Watch Instantly is about the here and now, and Netflix is priming to respond to that time frame."[40] Episodic entertainment with a strong hook to continue viewing emblematized the new mode of consumption. To be sure, Netflix had always foregrounded its subscription plans' supposedly limitless number of titles, a major marketing plank in its efforts to compete with Blockbuster. Given the fixed number of physical disks available to be held at home and the temporal constraints of the postal service, however, subscribers could only consume a de facto circumscribed number of films or episodes before having to wait for the arrival of new disks. What (UK) observers at the time called box-set culture—the behavior now universally termed binge-watching—first became possible with the new on-demand version of Netflix.

Thus, the arrival of streaming and the truly all-you-can-watch subscription plan ushered in a renewed core purpose for the recommender system: maximizing viewing time as a means of retaining customers. Recommendation was no longer, akin to critics' tips, a case of gesturing to the highest-quality film or series available. Rather, the task transformed into promoting content likely to occupy the individual viewer for as long as possible, babysitting him or her for such temporal lengths that the subscription would seem as indispensable as electricity or water. Under the DVD paradigm, the recommender system aimed to manage excess by helping overwhelmed or uninformed customers choose content beyond new releases, balance out bottlenecks in the physical inventory, and spotlight films that customers were likely to enjoy and for which Netflix had negotiated advantageous financial terms. Of course, even in those days the company had no desire to see disks languish on coffee tables for weeks on end; this could lead to perceptions of superfluous subscriptions, not to mention catalyzing dissatisfaction among other customers waiting for those particular DVDs to arrive. And yet Netflix never expected that any given customer would view more than five, ten, fifteen, heavy users perhaps twenty or thirty hours of content per month: the equivalent of three to twelve feature films. Under the new delivery regime, however, quantities of usage, what Netflix internally calls "engagement," became paramount.

The recommender system's model of cultural choice also transformed: from the critic's style of maximizer to the glutton's satisficer. A 2015 academic article by Netflix developers Carlos A. Gomez-Uribe and Neil Hunt explains the basic science and intentions behind the new Cinematch, drivers that pertained to new sorts of surplus management, taste, choice, and technology, and also to the company's financial health. The engineers make a case for engagement as Netflix's primary objective and the recommender system as the central tool for meeting this goal. Although, according to Gomez-Uribe and Hunt, "Internet TV"—note the term as a sign of the company's new self-understanding— "is about choice" and compares well to linear broadcasting in this respect, "humans are surprisingly bad at choosing between many options, quickly getting overwhelmed." The typical Netflix subscriber will only devote sixty to ninety seconds on average to choosing, skimming through ten to twenty titles (and examining only about three in any detail) on one or two screens of content. After this brief period of activity, the user will either begin watching something of interest or abandon the search as failed. "The recommender problem is to make sure that on those two screens each member in our diverse pool will find something compelling to view, and will understand why it might be of interest."[41] In addition, the developers write, the solution to this technical problem serves a much higher purpose: the company's economic value and growth potential. As the "core to our business," the "recommender system helps us win *moments of truth*," which the men describe as finding "something engaging within a few seconds, preventing abandonment of our service for an alternative entertainment option." To be sure, the system also helps the company match niche content to its intended audiences and thus create better value for the range of Netflix's acquisitions and overall catalog, a value that the company assesses with a metric they call effective catalog size (ECS). (The developers claim that their recommender system has improved ECS by a factor of four.) Most importantly, however, proper recommendations "lead to meaningful increases in overall engagement with the product (e.g., streaming hours) and lower subscription cancellation rates." According to the engineers, the new system dramatically decreased subscription cancellation percentages to the low single digits. Having observed a "strong" correlation between increases in viewers' streaming hours and customer retention, the developers estimate that "the combined effect of personalization and recommendations save us more than \$1B per year."[42]

These findings hastened the gradual move from a barebones collaborative filtering system—in which users were grouped into similar taste clusters—toward a more complex latent factorization model, by which both users *and* items were decomposed into "tastes spaces" in order to generate more precise recommendations. As Netflix shifted from DVD to streaming, from the temporality of days and weeks to instant video, its recommendation idiom transformed both in terms of potential sources (calculating viewing behavior rather than rented titles, online lookups, or ratings) as well as purpose: to find something to watch *now*. This shrunk the traditional lapse between recommendation and consumption. Unlike the newspaper review and the cinema visit (or even the television listing and program), suggestion, consultation, and starting to view transpired more or less in real time. In turn, the goal of surplus management shifted from sorting imbalances in the location of physical objects to alleviating the user's rapidly increased difficulty of choice. As we shall see, these developments altered the conception of taste at the heart of Netflix, eliciting and addressing implicit behaviors rather than user inputs. From regarded stars to disregarded thumbs up or down, from user-based collaborative filtering toward more item-based metadata: the new recommender system presumed that humans' explicit articulations of taste are unreliable and not entirely rational, concluding that observing actual viewing behaviors reveals more about preferences.

The remainder of this chapter highlights how the Netflix recommender system changed in the era of streaming. The discussion foregrounds three major shifts. First, the new technological approach to recommendation allowed for more descriptive explanations but also, by strategically veiling certain criteria, indulged users' taste self-identities. This process aimed to increase credibility. Second, the revamped system blended several different new recommendation sources and measures, including short- and long-term popularity, critical acclaim, and personalization. It thereby remediated several legacy forms, including criticism. Third, in a bid to counter a typical fault of algorithmic recommendation engines, "overfitting," and despite the company's headline rhetoric surrounding personalization and choice, the modern Cinematch works on the principle that the most useful form of personalization actually reduces personalization. It functions on the principle that many if not most human beings poorly understand and articulate their own tastes.

Because of Netflix's opaque and evolving approach to Cinematch, the goal here is not to *reveal* details of the recommender system in the sense of uncovering a truth hidden to outsiders, like investigative journalism. Nor do I intend to reverse engineer (as some have tried). Rather, the discussion highlights major structural features and the information that Netflix itself chooses to present externally. It scrutinizes how the platform explains and justifies its recommendations, analyzing the cultural implications of this presentation and establishing why groupings, classifications, explanations, and context—like in criticism—maintain a role in cueing consumption of films and series.

A. Engineering Credibility: Latent Taste Factors and Explanations

The Netflix Prize (2006–2009) and concurrent developments among the company's data engineers resulted in a few crucial technical changes in the nature and form of the recommender system. These included, above all, a blend of two underlying algorithm types, the most important of which is matrix factorization (also referred to by one of its applications, singular value decomposition, i.e., SVD).[43] In basic terms, matrix factorization and SVD refer to means by which an algorithm can predict certain unknowns in a data set (e.g., films that a user has never rated) by decomposing both users and films or series into a "latent set of factors." This type of predictive model can pinpoint tastes that a more basic collaborative filtering approach may miss. Rather than basing recommendations purely on what users with overlapping preferences appreciated, the technique, as AI expert Pedro Domingos explains, "first projects both users and movies into a lower-dimensional 'taste space' and recommends a movie if it's close to you in this space."[44] In other words, the algorithms, fed with Netflix's billion points of taste data, sort the films into categories that are "latent" or hidden, because not predetermined by programmers. Rather than instructing the computers to ferret out romantic comedy series or Woody Allen films, in other words, developers let the algorithms group the content by factors more specific than traditional genres and other common media content descriptions: amount of blood or explosions, type of humor (dark, satirical, slapstick), and so on. Unwittingly or not, the developers reverse-engineered an insight mulled over by the great and good critics, a feature evident in many word-of-mouth recommendations: traditional genre labels like "horror" or "comedy," while useful shorthand for identifying and broadly differentiating between audiovisual media, are too coarse to be predictive of taste. (If genre classifications *were* more predictive, there would probably be no need for most critical reviews, let alone advertising and word of mouth.) Even the most diehard fans do not love every horror film, because "horror" in fact encompasses a wide range of styles, attitudes, trajectories, and timbres, let alone stars, historical and cultural origins, and production staff. Decomposing content qualities and user tastes into more minute parts, "latent factors" of which many have no obvious name, yielded more finely tuned recommendations. The eventually winning Netflix Prize team deemed matrix factorization/SVD to be essential to the "most accurate" approaches to the problem.[45]

Using these algorithms in such constellations was part and parcel of a broader structural shift away from a chiefly user-based collaborative filtering model toward a more hybrid item-based collaborative filtering. That is, in order to predict what content a given user might enjoy, Netflix would henceforth pay more attention to the similarities between productions rather than solely to the similarities between users' tastes. In some ways, this change brought the company's recommender system back to its origins: after all, the very first iterations of Cinematch attempted to link films to users by offering recommendations with similar casts, crews, or genres of previously viewed titles. Netflix had abandoned these efforts, as such variables proved to be poor predictors of future tastes. The innovation of Netflix's post-2012 item-based collaborative filtering system would be a tagging and "microgenre" protocol. The company employed dozens of independent contractors, who worked ten to twenty hours per week assigning metadata to films and series based on a thirty-six-page training manual. These taggers collected some traditional forms of information (film locations), but also included seemingly more obscure reference points, such as the protagonist's job. According to the company, the procedure—by which the evaluator chooses from among one thousand different tags and scalar (1–5) values to describe the type of humor, the amount of romance, the nature of the narrative ending, the film's or series' tone, characters' attributes (including moral status), the level of gore, the "squirm factor" (awkward, cringe-inducing comedy), and so on—yields more effective outcomes than previous iterations. "People consume more hours of video and stick with the service longer when we use these tags," according to Yellin.[46]

Netflix users do not see any of these tags or scales on their interface. However, these metadata are instrumental for fitting content into Netflix's proprietary classification system of nearly eighty thousand "microgenres," which provide the headings for the content rows that make up the bulk of each homepage. For a 2014 *Atlantic* feature, Alexis C. Madrigal scraped data off the Netflix website to find 76,897 genres, and established patterns among them, including repeating semantical and syntactical forms. In other words, Netflix maintains its own vocabulary and grammar for describing genres, including a stable set of adjectives such as national origins and a substantial yet finite list of noun descriptors. Time periods and source material ("based on real life" or "based on classic literature") are also staples of the labels, which take on a regimented sequence (adjectives before region before noun genre before source material before setting before time period before subject matter): for instance, "Sentimental European Dramas from the 1970s."[47] The classification names overlap only partially with traditional genres, whose use has arisen organically over time, in a discursive consensus among industry, press, and audiences.

The new system offers further advantages over pure collaborative filtering in terms of credibility. The tagging procedure helps Netflix establish similarity between items and thus informs rows such as Because You Watched, which is based on a non-personalized algorithm and presents users titles that resemble those viewed previously (rather than those that a "similar" *user* watched).[48] The tags and microgenres enable Netflix to articulate, in the simplest terms, explanations: *why* any given content might be enjoyed by the user, to provide the recommendation with context, and thus increase the chance of take-up. For example, *Stranger Things* (2016–present) will have varying levels of appeal to any given user depending on the titles surrounding it graphically and whether the row is pitched under a headline containing the descriptors Science Fiction, Mind-Bending Scary, Coming-of-Age, Set in the 1980s, US Series, Netflix Originals, or Starring Winona Ryder. Context and descriptions form vital parts of the recommendation procedure. Indeed, Yellin revealed that the company aspired to put personalized suggestions into descriptive language that resembles genres—or more specifically: quirky independent-video-store "staff picks" shelves, those sometimes purposefully bizarre aggregations of Australian Westerns, Swedish Romances from the 1920s, Bollywood Horror from the 1970s presented by someone named Andy or Lucy—rather than a list of purely numerical quotients. As such, the cultural objectives of tagging (to establish qualitative, organic, Netflix-specific genres) serve grander economic aims: "Members connect with these [genre] rows so well that we measure an increase in member retention by placing the most tailored rows higher on the page instead of lower."[49]

Such programming decisions bespeak Netflix executives'—unconscious or conscious—understanding that recommendation *explanations* substantially affect users' consumption decisions. The new recommender system is sensitive to the fact that humans have a deep-seated need, well established in the fields of psychology and marketing but also in the computer science literature on the subject, to understand *why* they like something. In general, explanations make people more confident in recommendations, more trusting of the recommender, and more willing to accept suggestions of unexpected, serendipitous, or diverse items outside of their comfort zone.[50] The microgenres reduce perceptions of surplus and reduce the torture of choice by managing a long list of recommendations into easily decipherable headlines, descriptors that might better induce consumers to engage with a film or series. For example, I may be (slightly) more likely to try watching the film *Trolls* (2016), a production I would otherwise never countenance, if it were perhaps pitched as Mind-Bending Psychedelic Nostalgia and packaged together with *Koyaanisqatsi* (1982), rather than under the rubric of Children's, Animation, Dreamworks productions, or the (for-me) dreaded Anna Kendricks and Justin Timberlake.

There are nuances and exceptions to the general rule about explanation transparency. Moreover, debate abounds among computer scientists about how much transparency to provide and precisely how to justify suggestions.[51] Consumers are more likely to respond to a product contextualized in a way that confirms their *self-understanding* of their taste, and thus identity. If I think of myself as an Action-Adventure guy and definitely not a rom-com fan, I may be turned off by a list of Romantic Action-Adventure Movies— even if I would, were I to actually watch them, enjoy the recommended films. For this reason, Netflix engineers programmed Netflix's microgenres with intentional limits on transparency. Yellin describes how the recommender system reveals only certain aspects of the content's organizing principles and thus users' taste: a row listed as Action-Adventure Movies will be personalized to include only those exemplars with significant romantic subplots (if prior viewing behavior shows a predilection for higher-than-average levels of romance) and yet will not indicate this in the interface label. In other words, the row will read Action-Adventure Movies, rather than *Romantic* Action-Adventure Movies. As anticipated earlier in this book, users value certain qualities and types of justifications (those that mimic word of mouth or reference the system's past success in matching content to the user) more than others (critic's stars) or the mere quantity of information.[52]

Even Netflix's manner of presenting film and series information, in company lingo "Evidence," demonstrates how the system seeks to increase the credibility of its recommendations with personalized forms of explanations. An algorithm determines which information to include (or exclude) in the thumbnail and "Further Information" title page (e.g., synopsis, cast credits, awards, percentage of match, similar titles). Netflix personalizes which of several publicity images to present, based on viewer tastes: one user may see a still that emphasizes the content's romantic subplot, while another will be presented with a fire or explosion scene to represent that same film.[53] The platform's selective inclusion of award nominations and wins confirms what marketing professionals and sociologists of art have noted for years: Oscars, Golden Globes, critical distinctions, and film festival appearances (i.e., elite or insider legitimation), encourage some audiences to seek out and watch films and series. For others, however, this sort of messaging has the precise opposite effect, inhibiting their consumption because of their preconceptions that such productions will be boring, indecipherable, self-indulgent, politically self-righteous, or middlebrow.[54]

Whether guided by a masterplan or trial-and-error stumbling, Netflix's modern recommender system adheres to established dictates for legacy suggestion forms, as observed in empirical studies. We know, for example, that the presentation of, and language used to describe, cultural products crucially affect users' ultimate appreciation. Studies have shown how the description of and context for a cultural consumption choice are crucial for decision-making as well as perceived enjoyment, whether food, music, or films. People respond to the same avant-garde music differently when it is called "modern" or "popular" rather than "experimental" or "classical"; food menu language and wine labels have dramatic impacts on evaluations.[55] Furthermore, categories and the shapes of categories are decisive in terms of enjoyment; Netflix rows offer bite-sized pieces of taste descriptions, readily decipherable and consumable. "We like things more when they can be categorized," Tom Vanderbilt writes in his study of taste. "Our pattern-matching brains are primed to categorize the world, and we seem to like things the more they resemble what we think they should."[56] Although Netflix's headline rhetoric privileges data above all else, ultimately the company has created a recommender system with acute sensitivity to Bourdieu-esque conceptions of the relativity and cultural constructedness of taste. It remains careful, when shaping explanations, not to disturb users' self-regard in terms of their cultural preferences and social identity. Better to give the user romantic action films under the category Action-Adventure Movies, lest he or she reject the productions for questioning his or her self-image.

Netflix recommendation explanations are in general *why-explanations*. Some explicitly answer the question why the content will likely interest the viewer and why (on what basis) the content is being recommended: for example, Because You Watched (assumption: enjoyed) Film X or Series Y. Other elements of the interface (keyword descriptors after the percentage match like "dark—emotional—tearjerker," some microgenre adjectives, the text summary of the content, and trailer) also implicitly provide *how-explanations*, in other words, how the content will entertain the user (Mind-Bending Crime Dramas; Scary Cult Movies). Some rows use popularity as a justification, appealing to humans' tendencies toward social herding and bandwagon effects, and mimicking word of mouth: Trending Now; Popular. Moreover, the titles composing each row, no matter the names or content, are yoked together, and thus implicitly explained, by the rubric of *similarity*. Finally, recommended titles are assigned a percentage, the prediction of the potential for the content to please the user. On my Netflix, *Prime Suspect* (1991–2006) rates a "94%," whereas another detective series, *La Mante* (2017) scores only "82%." This figure inoculates the system against misses (I can tell myself that I must count among the 6 percent of like-minded users who do not like *Prime Suspect*), provides an appearance of scientific rigor (a "result" rather than a tip or hunch), and reminds the user that the recommender system is actually functioning. Together with the text, the groupings, rows, presentation, and overall design, the percentage provides what experts call a "fidelity claim," an assurance that the reasoning behind the recommendation is sound, "comparable to information about the expertise of a human advice giver (e.g., references to academic degrees and job titles)."[57]

Netflix's laconic, largely content- and case-based explanation style is of a piece with, and influenced by, its underlying recommender algorithms. Nevertheless, it represents a conscious choice: one of several options among the many that competitors deploy. The explanation style of Amazon's retail recommendation engine, for example, is collaborative-based: that is to say, it maintains a form of word of mouth based on user-user overlaps ("customers who bought this item also bought..."). The *New York Times'* Watching engine, in contrast, does not rely on extrapolating predictions from prior ratings or behaviors and instead deploys a conversational explanation style: users must respond to several questions about their mood and taste in order to receive suggestions. (Still other, much less successful and now defunct engines used demographic explanations: few people, apparently, want to be told that a certain film may appeal to them because they are a woman, Black, single, or have a high income.)[58]

Netflix's simple explanations and contextualizations gesture toward, if not outright fulfill, the basic definition of criticism: evaluation grounded in reasons.[59] To be sure, the words "Because You Watched *Taken*" (2008) and a scrollable procession of fifteen action thriller thumbnails would hardly seem to compare to a six-thousand-word Pauline Kael essay. But, with most media consumers unwilling to spend more than a minute making entertainment selections, computer scientists advise that explanations be simple and direct. In turn, the efficiency of selection represents another important criterion of explanation besides credibility, transparency, effectiveness, and persuasion.[60] A recommendation explanation need not be an exhaustive justification, but must "make [itself] clear by giving a detailed description," one textbook advises developers: "So, an explanation can be an item description that helps the user to understand the qualities of the item well enough to decide whether it is relevant to them or not."[61] I will return to the comparison between the relative use value of human-composed reviews and recommender systems, and to the feared threat to critics' vocation, in due course.

B. New Recommendation Sources and Framings

The switch from assessing users' stars to real-time measurement of their viewing behaviors ranks among the most significant turning points in the development of Netflix's recommender system. Even in its early days, Cinematch had measured members' ratings and noted previous rentals to determine future predictions and improve the personalization and effectiveness of its recommendations. It had also registered website searches and clicks and inferred that the more time the DVD remained with the user, the lower her or his motivation to watch it. Overall, however, the early system depended primarily on subscribers' self-supplied evaluations of their satisfaction with the content, the star ratings.

Indeed, some key clues about actual consumption were unavailable. The company could not definitively say whether a DVD languishing for a month at home was returned with such delay because the user had viewed it twenty times, could not bear to watch it, or had misplaced it under a stack of magazines. Even if a subscriber gave a film five stars, the recommender system could not independently determine whether he or she watched it three times in a row, abandoned it after five minutes, or even started it at all. Some members provided tens of thousands of ratings; others provided none. Indeed, the very utility of self-reported ratings was doubtful at best. A proportion of users tended to rate past choices in irregular bursts, misremembering their experiences and inflating or deflating their assessments in the fog of time. For some subscribers, three stars meant a very good viewing experience, if not an all-time favorite; for others, anything less than five stars denoted a dud.

In response to these challenges, Netflix changed recommendation sources and framings. Refining the system after the introduction of streaming video, the company could directly measure which content—and how much, how often, and over what time span—each subscriber actually watched. Netflix began to pay less heed to star ratings and eventually replaced them with a binary digit: thumb up, thumb down. Viewer behavior, rather than explicit expressions of taste, became the primary metric to determine preferences and generate future personalized selections. This process marked a further departure from the "best works" style of much criticism and an embrace of a more subtle and personalized word of mouth: flattering the user with headlines—if not necessarily content—that correspond to her or his taste self-image.

In recent years, Netflix now claims to incorporate an exhaustive set of data points for optimizing recommendations. These include members' ratings (i.e., thumbs, of which the company has billions); overall popularity (itself computed in many ways, in particular regions and time ranges); number and duration of streaming plays (also weighted for time of day and week and device type); items added to watchlists; content metadata (including genre, actors, directors, production year, parental rating); calculations of which presented content has translated into user clicks, scrolls, mouse-overs, or views (including time spent on a page); and searched-for titles—not to mention "many other features such as demographics, location, language, or temporal data that can be used in our predictive models." In order to process the data emerging from each of these sources, Netflix's claimed one thousand data engineers must program and refine dedicated algorithms gleaned from the laundry list of machine-learning applications.[62]

The results of this data processing emerge in real time on users' individualized homescreens. In this way, Netflix turns mere presentation into recommendation. Since the introduction of streaming, in the words of Amatriain and Basilico, "Everything is a Recommendation." More precisely, nearly every user action informs Netflix suggestions: "Personalization starts on our homepage, which consists of groups of videos arranged in horizontal rows. Each row has a title that conveys the intended meaningful connection between the videos in that group." Indeed, the rows themselves—Continue Watching, My List, Thrillers, Dramas with a Strong Female Lead—appear from top to bottom in a sequential arrangement that reflects what Netflix believes will stoke the user's interest and focus his or her attention.

Context, order, framing, access, and default options: research shows that all of these factors heavily affect choices, above and beyond supposedly innate preferences.[63] Indeed, rows and ranking count among the most important tools in Netflix's presentation of each user's personalized home screen, the recommender system's key visualization and interface. Various algorithms power various rows that feature varying degrees and types of personalization (based on prior behaviors and collaborative filtering) and popularity (or other global indicators). In turn, a whole host of algorithms support ranking functions.[64] The order of the categories and the sequence of titles within the categories reflect attempts to personalize and promote; titles and rows on the left-hand side and top, respectively, are more heavily recommended than those hidden before horizontal and vertical scrolling. According to Amatriain and Basilico, the user's lived experience of recommendation derives from these presentational tactics: "Most of our personalization is based on the way we select rows, how we determine what items to include in them, and in what order to place those items."[65] The system is designed to order, rank, sort, and filter content so as to "maximize member satisfaction and month-to-month subscription retention, which correlates well with maximizing consumption of video content. We therefore optimize our algorithms to give the highest scores to titles that a member is most likely to play and enjoy."[66]

In this effort, the Netflix interface seamlessly conflates recommendation forms of different provenances and purposes. For example, the row entitled

Top Selection (or, more recently, Top Picks), a product of Netflix's personalization algorithm (personalized video ranker, PVR)—according to Amatriain and Basilico "our best guess" at the titles the subscriber is most likely to watch—shares an aesthetic with the Trending Now row, a selection that relies more heavily on overall popularity. These rows are in turn visually indistinguishable from the many microgenres with headlines based on format or location (Series Set in London), narrative style or affect (Thrillers), or cultural status (Award-Winning Films). These latter microgenres feature, however, "some of the most recognizable personalization in our service."[67] And yet, it is important to note, the uniformity of these categories graphically veils the radically different sorts of algorithms behind them and more importantly the fundamentally different sorts of tastes they imply: great works (tastemaker-awarded distinction), social consensus (popularity), and personalization (whether based on genre, actor, or otherwise). These underlying taste types correspond to the historical recommendation function of critics, top-ten lists, and trusted friends' word of mouth, respectively. In recent years, the Netflix Originals promotional row has assumed, on some devices, an exceptional, markedly different visual style. Taller, grander, coherently branded, and richer in detail, the design suggests another genealogy: advertising.

Despite the heavy rhetorical focus on bespoke suggestions, in other words, the modern Netflix recommender system blends the traditions of legacy provider purposes: chief among them critical acclaim and short- and long-term popularity, besides and alongside personalization. Although in practice the size will differ according to the device and respective underlying catalogs of different geographical territories, the typical homepage contains approximately forty rows and up to seventy-five titles in each row.[68] Each row contains several dimensions, and varying levels, of personalization: the fact that the genre is even shown to the user; the position of the row, with the most personalized rows higher up on the page; which subset of titles to show within that row; and in which order the titles will appear, with the most tailored films or series appearing on the left side of the row. The Top Picks row uses an algorithm that attempts to find the most personalized recommendations across the catalog, without regard to genre categories. Popular, in contrast, presents an algorithmic selection by which popularity weighs more heavily than personalization. Trending Now—a row constructed after Netflix noticed that short-term temporal trends effectively predicted user viewing behaviors—offers films or series that have been popular in the last few minutes and days, a row that can also automatically generate seasonal recommendations for Christmas or romantic films on Valentine's Day. It can also profile short-term interest in disaster movies or documentaries during hurricanes or tumultuous political events, one of the reasons that *Outbreak* (1995) and *Contagion* (2011) featured so prominently in the early days of the 2020 COVID-19 pandemic. Even these rows, however, and even the Continue Watching and Watch Again rows, feature personalization in their title choices and rankings.[69]

The mélange of recommendation types responds to a discovery that Netflix's and other data engineers have made in developing algorithmic systems: pure personalization—despite its success as a branding tool—produces unsatisfying outcomes. Merely recommending, day after day, a user's favorite genre, style, humor, director, or level of violence or romance cannot effectively sustain members' attention and retain subscriptions.

C. You Don't Always Want What You Want: Ensuring Freshness with Non-personalized Personalization

Indeed, beyond personalization, critical acclaim, and popularity, a fourth major criterion remains crucial to furnishing successful recommendations: "freshness," Netflix developers' term for the diversity, serendipity, and novelty of suggestions.[70] In other words, even if you largely watch police procedural series and political documentaries, the system will bet that there are times when you would rather see something else and introduce other types of content accordingly. Gomez-Uribe and Hunt add that their personalized ranking algorithms function "better when we blend personalized signals with a pretty healthy dose of (unpersonalized) popularity."[71] Indeed, each of the homescreen categories beyond the genre rows maintains varying levels of personalization, diversity, freshness (including both personal and global temporal trends), and general popularity.

The conceit behind Netflix's recommender system bespeaks that old newmedia conviction that data do it better: algorithms can more ably understand tastes because they remain unswayed by cognitive and cultural biases, groupthink and herding, illusory correlations and hot-hand fallacies, superstition and conjecture: in short, all of the flaws of human interpretation created by social acculturation.[72] At the same time, data engineers who refine Cinematch and similar products must actively combat one of the perennial and most intractable bugs in machine learning: "overfitting." Domingos supplies an anecdote to explain the concept: a young white US-American girl who sees a Latina baby and exclaims "look, Mom, a baby maid!"[73] In this case, the girl's insufficiently wide interactions with Hispanics—limited to her own housecleaner—has led her to the false conclusion that all Latinas are maids. Overfitting is essentially a misinterpretation that results from overinterpreting insufficient input information. Returning to recommender systems, it refers to problematic coding that gives certain data too much weight. By attempting to find grounded reasons for watching or liking certain content, an overfitting recommendation engine excessively narrows the selection presented to the user. It is a form of overpersonalization.

Some commentators see overfitting, and a propensity to filter bubbles and cultural echo chambers, as inevitable problems of algorithmic recommender systems. This thought, we shall see subsequently, remains prevalent among some lay users as well. Nevertheless, programmers have seriously debated overfitting at least since the mid-1990s in the attempt to escape this pitfall. Over the years, they have developed numerous techniques to include unexpected items and inject novelty, serendipity, and randomness into personalized recommendations.[74] Overfitting can be caused or made worse by random events, incorrect inputs (my niece used my Netflix profile), and too much "noise" in the data. A major newspaper ranking *The Godfather* (1972) as the best film of all time might create a huge short-term spike of viewers; however, this should not lead the system to believe that each of these users is suddenly forever interested in 1970s Italo-American gangster film adaptations. As Domingos writes, well-designed recommender systems seek to avoid overfitting by stopping short of attempting to perfectly match predictions to existing data. Allowing for some natural flexibility by aiming for a high probability, rather than a perfect match, will present some clunkers to the user and yet enable some discovery. Such a procedure will also escape the risk of reducing suggestions to content the user would definitely enjoy but likely has already seen.[75]

The computational overfitting problem gestures to a central paradox of recommender systems in particular and taste in general: the best personalization is not too personalized. In extensive testing on their users, Netflix engineers found that providing users with increasingly specific recommendations based on star ratings, prior viewing behavior, or any other cue—treating users' tastes as puzzles to definitively resolve—did not lead to greater satisfaction. Amatriain and Basilico report how they used item popularity as a baseline to determine the efficacy of the recommender system. They measured the company's new-media conviction and personalization branding against the fact that popularity was in itself a powerful predictor, since "on average, a member is most likely to watch what most others are watching. However, popularity is the opposite of personalization."[76]

Here too the recommender system's development followed observations that have long been known from empirical studies of taste, and indeed will reappear in the original audience studies I undertook for chapter 5. Humans—unlike computers—are *not* immune to social herding; many if not most enjoy watching what others like to watch. According to Amatriain and Basilico, using "predicted ratings on their own as a ranking function can lead to items that are too niche or unfamiliar being recommended, and can exclude items that the member would want to watch even though they may not rate them highly." People, in other words, want to watch what other people are talking about, whether or not these films or series map on to their most preferred microgenres or correspond perfectly to their optimal levels of romance or action. Humans are not efficiency- or utility-maximizing machines.

The developers explain how they endeavored to square the circle of personalization versus popularity. Rather than using either popularity or predicted ratings by themselves, they sought an aggregated matrix that required many data sources and included various rows that balance pure personalization with popularity, freshness (new titles), and a measure of diversity.[77] The Netflix recommender system thus represents a finely tuned negotiation between personalization and popularity, between fitting the user's viewing patterns precisely and avoiding overfitting by introducing a measure of discovery and diversity. To this end, a few key functions—presentation, ranking, and item-item similarity—have become essential to maintaining an equilibrium. So too do rows such as Popular and Trending Now attempt to compensate for more personalized rows such as Top Picks. Paradoxically—and despite the company's rabid personalization rhetoric, its ideological diatribes against editorial content and linear broadcast scheduling—the Netflix recommender system maintains some key features that simulate general-interest newspapers, broad-church multiplex cinemas, or network-television broadcasting.

In sum, in order to combat overfitting, Netflix recommends both what it expects individual users to like as well as what it knows that subscribers, in aggregate, appreciate. The recommendation engine combines personalization with some of its opposite: overall popularity and diversity. Although counterintuitive in the face of Negroponte's predictions and business mavens' exhortations, the most effective recommender systems—from the perspective of Netflix and other commercial providers who seek to retain subscribers and keep them watching as much content as possible—do not attempt to deliver content based on users' strongest predilections. Company engineers not only take into account the fact that a human's taste varies and changes. They presume her or his taste includes products, styles, genres, stars, and narrative forms that she or he has never encountered and which may not yet exist. Even if Netflix were somehow able to tag every second of every title across an infinite number of characteristics to feed into an infinite number of microgenres and record even more aspects of viewers' behavior (eye-tracking attention, for example), this would still only indicate past behavior. Even if Netflix could precisely chart users' preferences in films and series to the finest detail—like mapping the human genome, or the empire in Borges's "On Exactitude in Science"—feeding this precise picture of taste back to them would likely *change* their preferences: too much of one thing, even a favorite thing, causes sensory satiety.[78] Like the ebbs and flows of stock markets, bond prices, and currency exchange rates, in matters of taste past behavior is no failsafe guide to future performance.

Another key reason for the apparent paradox of personalization and popularity has to do with the incongruity between users' claimed preferences and their actual viewing behaviors. In 2017, Netflix replaced its star ratings with a simple thumbs-up or thumbs-down button. The symbol recalled the old Gene Siskel and Roger Ebert rating system, much derided by other film critics at the time for the way that it seemed to essentialize complex evaluations into fickle and snap Nero-esque judgments.[79] Furthermore, the iconography resonated with Facebook's and other social media companies' "like" buttons, a similarity that invoked pronouncements of a "tinderization of feeling," in reference to the contemporary hook-up app. In the *New York Times*, Tom Vanderbilt commented that the thumbs represented a "kind of dumbing down" of complex evaluation into thoughtless binary judgments and, rehearsing the filter-bubble argument, claimed that they would only "steer us deeper into our own proclivities."[80]

To be sure, Netflix's stated purposes were far more pragmatic. According to Yellin, the star-rating system was statistically imprecise because of its five-scale range: a two- or three-star rating might have many fine-tuned motivations, but compared to a simple binary, it remained ultimately unhelpful in refining recommendations. The Netflix Prize had a bevy of algorithms correcting for different base-line ratings among viewers but also within any individual viewer's ratings.[81] It demonstrated that at "both the individual and aggregate level ... Netflix stars are far from fixed. Rather, they are like free markets: prone to corrections, bubbles, hedges, inflation, and other forms of statistical noise."[82] Netflix claimed that ratings failed as a robust indicator for personalized recommendation in the same way that identity-based classifications such as gender or geographical location had; they were not statistically insignificant or useless, but not sufficiently predictive either.[83]

The introduction of thumbs continued a series of unpublicized alterations to the recommender system's personalization algorithms. Each of these changes marked a trend away from trusting viewers' own ideas about their taste (as articulated, for example, in star ratings) and toward an accounting of preferences based on observed behavior. Confirming empirically (and no doubt unwittingly) the long-standing conclusions of Bourdieu and Thorstein Veblen, Netflix data determined a disjuncture between subscribers' ratings (what they said they enjoyed watching) and their actual behavior when streaming in real time (what they spent the most time watching). Stars, outward-facing expressions of taste, constituted distortions and enabled aspirational virtue signaling. Earnest features, issue-oriented documentaries, canonized classics, and critically acclaimed films—productions that some if not many viewers found a moral obligation to like, whether because of the film's subject or politics, or because of its legitimated status by respected tastemakers—were especially prone to such biases. Netflix data revealed that in the days of DVD, Al Gore's global-warming documentary *An Inconvenient Truth* (2006) hung around (presumably unwatched) in subscribers' homes for the longest periods.[84] Before the adoption of thumbs, programmers had to develop algorithms to selectively rebalance star ratings to take such biases into account.

The development of the post-2012, post-star Cinematch leaves us with three final insights about Netflix but also the larger taste mechanics of rec-ommendation. First, developers of recommender systems have gradually learned an uncomfortable fact: humans are not necessarily experts of their own taste. Many of us find it difficult to precisely articulate the contours of our own preferences in films and series, whether in the form of subject matter, traditional genre labels, or more abstract indicators such as levels of romance in action-adventure movies. As strange as it may seem, we are not always the best judges or succinct explainers of our own tastes. The Net-flix recommender system relies on this ignorance—and users' reluctance to learn more about films and series via other sources—in order to add value. It bets on non-expert users who do not (care to) understand subtle connections between cultural products, who do not (care to) understand film and television history, who do not (care to) furnish such cues from ads, phoning a friend, or reading reviews. (Examples of these phenomena abound in the empirical audience study in chapter 5.) After all, recommender systems, computer science textbooks remind learning developers, are geared to non-experts, "individuals who lack the sufficient personal experience or competence to evaluate the potentially overwhelming number of alternative items."[85] Those expert users—whether critics, academics, professionals, or hobby cinephiles and series junkies—have other, more direct means of accessing content, beginning with the search functions of various audiovisual media providers.

Despite impassioned new-media rhetoric regarding personalization, unfettered access, and free will, the Netflix recommender system operates under the assumption that users' choices are not fully rational. The notion of bounded rationality, long researched by cognitive scientists and behavioral economists, dictates that human limits, such as the brain's information-processing capacity or our finite attention spans, circumscribe our ability to act in entirely predictable ways.[86] Indeed, Netflix engineers stress that they regard customers' overt wishes with productive suspicion. Guided by viewers' behavioral data rather than stated preferences, the company largely ignores what customers say they want and instead provides them with what it thinks they need. According to Gomez-Uribe and Hunt, "what customers ask for (as much choice as possible, comprehensive search and navigation tools, and more) and what actually works (a few compelling choices simply presented) are very different."[87] This procedure strips out some of the virtue signaling and other social taste biases observed by Veblen and Bourdieu. This protocol also addresses a major conundrum of recommendation science: although humans like the abstract *idea* of choice—whether restaurant menu items, supermarket ice cream flavors, health care options, pension plans, television channels, or film titles—truly comprehensive selections overwhelm.[88] Netflix's contemporary menu interface provides an illusion of plenty and manageability at the same time. Its Watchlist feature manages surplus and defers difficult choices to later dates, thus inscribing a future viewing time and extending the projected value of the subscription.[89] De-emphasizing user-generated articulations of taste, Netflix pursues a trust-but-verify strategy of suggestion.

Second, the development of the Netflix recommender system complicates much research on traditional cultural mediation and taste for audiovisual products. Beyond tiny coteries of exclusive horror fans or chick flick aficiona-dos, humans are complex creatures with shifting alliances. We are willing and able to appreciate different content types on account of various underlying factors: a sci-fi film with a romantic subplot, a political thriller that appeals to fans of police procedurals because of narrative complexity, a taste for deli-ciously flawed and cynical characters that obtains whether the plot plays in the Wild West, Victorian England, or a dystopian future. In this way, the history of the recommender system reveals that traditional broad literary and industrial genres, while useful communicative shortcuts, are not always the most important vectors of taste; humans may be more omnivorous than many cultural observers and industry executives assume. The notion of the cultural omnivore has been part of scholarly literature for decades.[90] Yet many applied studies of taste in film and television studies, such as those examining *Star Wars* or *Star Trek* superfans, often rely on schematic or rigid understandings of preference. Variety, novelty, and serendipity—but also social bandwagon effects that make people want to watch what others are watching—remain essential components of cultural mediation, as important as deciphering a supposedly innate, unique, and exogenous taste. Netflix's decision in 2020 to begin sending emails and graphically highlighting on the homescreen a "Top 10 in [the user's country] Today" row with large numerals bespeaks this importance to users. Despite its personalization branding, after years of A/B testing Netflix has come around to the insight that the venerable box-office list—with its deeply social, rather than individualized, measure of taste—is an important motivator for people to select films and series. The audience study in chapter 5 will explore, test, and deepen this insight with empirical examples, showing how even in the supposedly atomized digital age, "keeping up" is a crucial driver for film and series consumption.

Third and finally, this chapter has narrated how the remediation processes of Netflix's recommender system have shifted over time, even while remediation as an overall strategy has persisted. Scholars have remarked how the company's perceived competition and purpose have transformed to match the company's intervention into different distribution and production markets. At various junctures over the years, publicity rhetoric has asserted the platform to be the "world's largest online DVD rental service," at others "the world's leading Internet subscription service for enjoying TV shows and movies," and at still others a "global TV network." In terms of its function and service, Netflix has been termed a video shop, television network, movie studio, distributor, or tech company. In turn, its interface has consciously evoked video-store aisles (assemblages of thumbnails arranged to recall DVD boxes) or a celluloid strip (its more recent genre rows). According to Ramon Lobato, this techno-aesthetic development seeks to "discursively reposition the site within the pantheon of older media technologies by moving the idea of Netflix away from video-store and DVD culture—surely a fading memory for most of its users—and realigning the service with that most resilient medium, cinema."[91] Each of these marketing and design efforts has depended on a peculiar and yet compelling paradox: highlighting the familiar (legacy forms, formats, and channels) and simultaneously claiming to innovate, disrupt, and even revolutionize. It has been the task of this chapter to demonstrate that this strategy is not only immanent at the level of business plans or platform aesthetics: the instruments and functioning of the recommender system, its mechanics of taste, also participate in this remediation.

FOUR

Unpacking Netflix's Myth of Big Data

ACCORDING TO THE PREVAILING VIEW, VOD recommender systems are catalyzing decisive and profound cultural effects. For these vocal objectors, the tools reorient cultural authority from humans to machines, a rebalancing that effectively evacuates audiovisual media selection, suggestion, and criticism from the stable of humanistic undertakings. Up to now I have hinted at these arguments; let us scrutinize them more closely.

In writings on the subject there exists a surprising overlap between tech cheerleaders and their academic critics. On the one hand, business gurus and the IT crowd celebrate the replacement of "editorial content" (human-generated words, subjective ideas, and biased assumptions) and pesky video-clerk word of mouth. According to this ballyhoo, the "video store with the top-rental list is mostly gone, replaced by the Netflix recommendation engine, one of the most sophisticated 'computers' ever built. This engine literally processes billions of data points across tens of millions of customers to maximize your entertainment experience."[1]

Many academics, on the other hand, agree that the service has revolutionary, culture-shifting effects, but they evaluate these developments in much less flattering terms. They take issue with Netflix's pretensions to personalization, gesturing to the murky ramifications of acting on mass-scale quantities of silently collected viewing data. In particular, commentators stress how these new qualities of recommendation compromise the integrity of classical forms of suggestion, especially criticism. Disputing the company line about consumer empowerment and connectivity, Sarah Arnold writes that Netflix's abandonment of user-inputted stars to instead surveil clicks, scroll-overs, and views, "represents a move from a measurement model that understands audience identity as culturally produced (and brought to the viewing experience) to audience identity as produced through data (and defined by data algorithms)." The quantification and "datafication" of taste, Arnold suggests, reduces audience agency and "masks more profound forms of individual manipulation and governance manufactured through data algorithms used by online platforms like Netflix."[2] She and other scholars suggest that by conducting such passive means of data collection (rather than soliciting active user feedback), "the Netflix model of measurement and prediction effaces the context, experiences, and identities of its users even more so than traditional measurement systems. The type of knowledge produced by Netflix works to negate the sense of a public, of a socially shared experience and of human agency. . . . The Netflix user becomes classified as a set of data," beyond the pale of "interpretation, judgment, and analysis."[3] This sort of argument has been advanced by several academics, who maintain that algorithmic recommender systems are a priori reductive and de-humanizing.[4]

Much in this vein, Neta Alexander seeks to highlight the "dangers and prospects of shaping cultural preferences based on methods such as data mining and collaborative filtering," questioning whether taste can "be translated into an empirical, mathematically based formula" and whether algorithms could replace cultural experts.[5] Noting the Netflix recommender system's increasing sophistication, she deems the engine no longer a simple matchmaker but now a "system that constantly translates seemingly chaotic behavior into recurring and therefore predictable patterns."[6] Exposing a mendacious discourse of personalization, customization, and choice, Alexander diagnoses a reductive (because algorithmic, quantitative) approximation of taste that hides a lack of quality in the platform's constantly changing catalog.[7] The "algebraic equation of taste" yields a feedback loop: "by documenting viewing habits and consumption patterns, they gradually change these very same activities." Sharp increases in series as a share of Netflix's offerings, after the platform's origins as a cinephile (i.e., predominantly feature-film) DVD service, evidence this phenomenon. Alexander concludes that "the distinction between the 'expert' and the 'consumer' collapses" because taste authority is reassigned to big data and algorithms.[8]

For his part, Daniel Smith-Rowsey ventures that the form of Netflix's data-driven personalization—in particular the (non)transparency of its micro-tag genre labels—"rests uneasily between both capitalist imperatives and hierarchies produced by Netflix's somewhat Bourdieuan authority." In its "re-definition of genre" including nineteen umbrella categories (horror, action and adventure), ca. four hundred subcategories (Italian Horror, Deadly

Disasters), and nearly eighty thousand microgenres (Visually Striking Father-Son Movies), the service has created an "intentional instability." Studios and networks "can, to a limited extent, pay to have films/shows pop up where they should not be," and yet "Netflix refuses to let users understand this process" and "excludes them from curating their Netflix 'recommendations' experience." Smith-Rowsey predicts that traditional genre markers, developed organically over time by viewing, criticism, and production practices, may disappear as "Netflix in effect privileges some films and shows and types of viewership, and to some degree reconstitutes what Netflix's sixty million users *think* when they think of film and TV."[9] The implication of Smith-Rowsey's argument is that—rather than an automatic effacement of traditional categories—in fact Netflix is heavy-handedly creating a new critical and cultural authority with its own idiosyncratic system of classification.

Finally, Ted Striphas rehearses the Netflix story as a milestone in the wholesale sell-out of the traditional humanistic values, ushering in a new "posthuman" era: "algorithmic culture." Since the 1980s, Striphas writes, "human beings have been delegating the work of culture—the sorting, classifying and hierarchizing of people, places, objects and ideas—increasingly to computation processes."[10] Furthermore, Striphas envisions a world in which the terms ("collective intelligence") and components (algorithms) of culture are held as proprietary by the organizations that develop them, and thus remain inaccessible to users. Algorithms seek ultimately to "conceal," Striphas argues, and, as "'socio-technical assemblages' joining together the human and the nonhuman, the cultural and the computational," they anticipate the "automation of cultural decision-making processes, taking the latter significantly out of people's hands."[11]

For Striphas and his collaborator Blake Hallinan, the Netflix recommender system represents a computational reinterpretation of basic humanistic assumptions: *culture à la technologie.* As a result of this appropriation, "questions of cultural authority are being displaced significantly into the realm of technique and engineering, where individuals with no obvious connection to a particular facet of the cultural field (i.e., media) are developing frameworks with which to reconcile those difficult questions," whereby "issues of quality or hierarchy get transposed into matters of fit."[12] In this "posthuman" scenario, "engineers—or their algorithms—become important arbiters of culture, much like art, film, and literary critics," to highly uncertain, and reductive, ends.[13] If experts' film reviews offered written evidence, context, and arguments for a numerical star rating—as well as the authority of the institutional source and critic's reputation—"the Netflix rating system reduces the opportunities for explicit context but continues to draw on implicit contextual information in the form of previous ratings and, more recently, other user data such as location and device."[14] Hallinan and Striphas foresee a world in which critics and humanities scholars may be no longer able to interrogate "computationally-intensive forms of identification and discrimination that may be operating in the deep background of people's lives, forms whose underlying mathematical principles far exceed a reasonable degree of technical competency," not to mention the ways in which cultural products are "optimized" in order "to ensure a more favorable reception, both by human audiences and by algorithms."[15] Striphas goes on to argue that the Netflix recommendation regime is fundamentally elitist, in step with an "authoritarian" concept of culture.[16] Despite Netflix's "populist rhetoric" about trying to "connect people to movies they love," its cultural mediation implies an Arnoldian "apostolic vision for culture." Collaborative filtering recommendation algorithms represent a pseudo-achievement of democratic legitimation: "Now anyone with an Internet connection gets to have a role in determining 'the best of what has been thought and said'!"[17]

In general, scholars' commentary on Netflix's recommender system and its algorithmic personalization functions revolves around the datafication or mathematization (and thus reduction) of taste, the recommendation engine's lack of transparency to users, and the challenge to traditional forms such as human-composed critical reviews.

I have sympathy with the sentiment behind these objections, which are not to be taken lightly. Nevertheless, it is important to nuance such claims: both Netflix's proponents and critics, usually not taking into account legacy recommendation modes, accept the premise that the recommender system's form and design (and thus effects) are fundamentally new. To be sure, the two sides' normative conclusions differ starkly: alternately, the system constitutes an innovative panacea to the limits and biases of the human mind, or a corruption of humanist criticism, taste, and culture. Yet both advocates (willingly) and detractors (no doubt unwittingly) magnify the company's data-collection and recommendation capabilities; both arguments amplify the importance, credibility, and use of the recommender system, ascribing to it an essentially novel and all-knowing quality. This chapter will dissect this myth, which Netflix and its enablers in the tech and business press actively seek to cultivate, and which even many of the company's critics and detractors implicitly accept. Analyzing scholarly, business, and tech discourse, company press releases and publicity, the discussion reveals how Netflix offers a performance of scientific objectivity, innovation, and differentiation as a way to establish the credibility of its recommendations and overall service—similar to the manner in which film critics must perform authority, knowledge, distance to (or familiarity with) the industry in order to establish their trustworthiness. This analysis will flow logically into chapter 5, whose quantitative and qualitative study of real users shows that most people have an at best ambivalent, circumscribed estimation of, and trust in, Netflix and other algorithmic VOD recommender systems. These systems also remain less widely used than assumed. These facts put scholars' pronouncements about posthuman culture-shifting effects further into question.

WIZARD OF OZ: THE MYTHOLOGY OF BIG DATA

We need to be skeptical of language that inflates, mystifies, and monumentalizes the recommender system, the big data that supports it, and the algorithms that guide it. In 2011, media scholars danah boyd and Kate Crawford published a think piece that defined and critically scrutinized the "mythology of big data" as "the widespread belief that large data sets offer a higher form of intelligence and knowledge that can generate insights that were previously impossible, with the aura of truth, objectivity, and accuracy."[18] Netflix actively seeks to cultivate such myths; the news media and (even critical) commentators have their own motivations to recycle and inflate them. This discourse preys on the widespread ignorance about back-end computer programming and ultimately serves as publicity that glorifies the company as cutting edge and indispensable, a blank slate for new-media-fantasy projections of all sorts.

As media scholar Tarleton Gillespie notes, public descriptions of how algorithms work "may seem like a clear explanation of a behind-the-scenes process," but may in fact be "'performed backstage,'" a discourse "carefully crafted to further legitimize the process and its results."[19] The story of the Netflix Prize, a science contest the company sponsored between 2006 and 2009, reveals the algorithmic recommender system's alchemy of technology, customer service, product differentiation, and publicity-seeking. The exercise in improving the accuracy of personalized suggestions constituted above all an effort to raise the profile of the company and to enhance the credibility of the recommender system as cutting-edge, objective, and scientific, a superior digital-age remediation of traditional cultural taste.

The competition inspired approximately ten thousand entrants—including engineers and computer programmers, mathematicians and physicists, philosophers and psychologists, hobbyists, Ivy-League scholars, and blue-chip technology firms' salarymen—whom Netflix provided with a data set containing a wide selection of real users' anonymized ratings. By 2006, Cinematch claimed to be able to predict subscribers' star ratings within one half of one star. The goal of the contest was to best this mark—and thus buttress Netflix's ability to predict which content individual users would enjoy—by at least 10 percent. The company offered to award $1 million to the winning team of developers.

From the beginning, the Netflix Prize was a widely discussed stunt. Unabashedly ambitious (if not nakedly delusional), CEO Reed Hastings hoped to mimic the Longitude Rewards, the early eighteenth-century government inducements for scientists to improve the calculations of longitude at sea. Megalomaniac chatter compared the task to Alan Turing's cracking the Nazi Enigma Code; military metaphors abounded.[20] Preceding the launch of the competition on 6 October 2006, Netflix's marketing team ramped up promotion with press releases and international media outlets covered the story as a major event. By 2 October, the *New York Times* had already issued a major feature on the contest, replete with laudatory quotations from computer science professors and a fawning platform for Hastings to emphasize the democratic aspect of his largesse: "Mr. Hastings said the Netflix prize was different from some others in that it required a minimal financial investment to compete. 'This will be one of the largest truly open prizes that's ever been done,' he said. 'All you need is a PC and some great insight.'"[21] The *New York Times, Wired,* and other major popular publications would regularly report on the prize and the contestants' progress up to and after the end of the competition on 21 September 2009, providing Netflix with ample free publicity. Business analysts have argued that the mastery of public attention rehearsed in the course of the Netflix Prize— and the award ceremony's studied hoopla of "branded banners, eye-catching props, and high-tech hardware"—later served as the model for the hitherto understated company's "barnstorming rollouts" of its various international services, beginning with Canada in 2010.[22] Indeed, the competition provided a boon for the credibility-building goals of the recommender system, not to mention brand awareness and subscription numbers. Although some insights derived from the Netflix Prize contestants were slotted into the current recommendation engine, it is telling that the company balked at fully incorporating the winning team's algorithms.[23] The publicity aspects of the competition, a bargain for the $1 million outlay, proved more valuable than the prize-winning algorithms.

The subsequent move from DVD to streaming, while not without its critics, on the whole ushered in a hitherto unprecedented glorification of the recommender system. With adulation usually reserved for popes in *L'Osservatore Romano*, the tech press shifted its full weight into the flourishes of futurology and science fiction that we know from Mosco's archetypes of new-media discourse. The system that would emerge by 2012 was "so sophisticated," one commentator breathlessly intoned, "that it could read peoples' movie tastes from behavioral clues and no longer needed much input from the ratings system." Hitting the "holy grail of customer feedback in the streaming service" and its ability to collect real-time data on when, how often, and how quickly users watch certain types of content, what they skipped, paused, or abandoned, Netflix now had a "richer and more personal . . . analysis of human behavior" than any focus group or interview. "If the algorithm chooses hits more often than misses, it captures the essential ingredient for a successful brand—our trust." A movie deliverer was ushering in the final triumph of the quantitative over the qualitative.[24] Such representations of an all-seeing, all-knowing taste expert, illustrated in fawning terms in regular *New York Times* portraits and *New Yorker* think pieces, gifted ample, uncritical space for Netflix's marketing messages about the power and mystery of its recommender system. "There's a whole lot of Ph.D.-level math and statistics involved," Netflix developer Neil Hunt boasted unironically in one such puff piece.[25]

A cast of sycophants from the tech, business, and media pages exalted Netflix as the revolutionary and unparalleled next big thing. Joe Weinman, the self-described "leading global authority on cloud economics" and author of the business handbook *Digital Disciplines*, uses Netflix as a key case study in his investigation of leading enterprises that utilize big data and other key technological developments in innovative ways. Noting that Netflix has "disrupted" the traditional video entertainment model in five ways (DVDs by post rather than video-store trips; a subscription rate rather than premium pricing for new releases; online browsing and queue; a recommender system, critical to revenue, profitability, customer retention, and customer lifetime value; streaming video), Weinman delivers a slack-jawed panegyric to the company's "accelerated innovation" and "solution leadership." Netflix, according to Weinman, is a "deadly competitor" because it "excels in virtually all of the digital disciplines," especially "collective intimacy," in other words, using big data to personalize content via the recommender system.[26] Such talk resonates with the early digital utopians, who predicted that new technological forms would empower consumers. According to these dreamers, VOD services, with their large content libraries, enhanced access and search functions, and data-driven recommendations, would create a "radical change in consumer attitudes" and a decidedly "active audience."[27] The increased abilities to collect and create data and metadata would allow for a "far greater understanding of audience demand, as well as providing for a more effective means of categorisation and archiving."[28]

In the gushes of Joe Nocera in the *New York Times* or Ken Auletta in the *New Yorker*, Netflix features as an apple-pie American pillar of innovation and entrepreneurialism, an omniscient builder of the best of all possible entertainment worlds. "Although news coverage now tends to focus on its shows," Nocera confides to his readers, "Netflix remains every bit as much an engineering company as it is a content company." Netflix, according to Auletta, represents a "digital revolution" of television.[29] Part of this mythology seeps into (or even more probably: emerges from) the utterances of business investors eager to prop up the publicly traded company's stock price. In a memo, one hedge fund manager wrote that although on the surface the company might seem to be a "massive video store," below "the surface, Netflix is akin to a think tank, creating algorithms to maximize the long-term value of each customer that it enlists." It is the latter, "unseen aspects of Netflix's business model, and its long head start, which differentiate it from the competition and may allow Netflix to retain the leading position in the industry for some time."[30] Overall, the reportage essentially replicates the company's PR messages. The "key thing that sets Netflix apart," a letter to shareholders states, "is our algorithms for personalization, which help members find content they will enjoy, and lead to increased viewing for any given set of content. Compared with simplistic 'most popular' merchandising, our algorithms add much enjoyment to our members' experience." It continues: "Our rate of learning is faster than any competitor because we have a larger membership from which to learn."[31] At heart neither a film distributor nor series producer, Netflix, as executive Ted Sarandos has said in interviews, is "a Silicon Valley-based intellectual property company that was born on the Internet."[32] Such comments resound with familiar tropes about large data-driven American companies in the entertainment business. Amazon is hardly a bookstore, another commentator opines; books were simply "Amazon's 'customer-acquisition strategy.'"[33]

Of course, the kernel of these comments yields some truth: yes, the search functions of Google and the networking, news feed, and photo-sharing aspects of Facebook all ultimately serve the goal of consumer data collection and pinpointed advertising. Nevertheless, the tone of press accounts is often less revelatory or critical than admiring and laudatory. The new-media mythologization of Netflix harnesses those identifiable beliefs and rhetorical forms pervasive among computer engineers, Silicon Valley managers, and Wall Street financiers. The result mixes blind belief in the free market, start-up entrepreneurialism, and the incorruptible, value-free perfection of science, with a vulgar (and masculinist) Darwinism, (Ayn) Randism, and an unquenchable thirst for data.[34]

In his polemical, revisionist book on the new-media giants of film and series, including Netflix, Amazon, Apple, and YouTube/Google, Michael Wolff takes such accounts to task, portraying the companies as arrogant, self-anointed "disruptors" who see themselves locked in a zero-sum game with self-doubting legacy media.[35] With groupthink gripping shareholders and press alike, Wolff describes a "revolutionary conviction" that "swept through the media industry: the new was certain and inevitable." A new reportorial class, of "pronounced and accepted bias, came into being: technology journalists whose very jobs and identity were hitched to proselytizing for technology."[36] Indeed, *Wired, Wall Street Journal*, and their competitors had vested interests in promoting such companies using the rhetoric of innovation and disruption. For Wolff, at least, this sort of evangelicalism belied a fundamentally conservative company, an "unrevolution," for its Negroponte-esque new-media rhetoric vastly outpaced the practical reality: Netflix was simply another television channel. "Other than being delivered via IP, Netflix had almost nothing to do with the conventions of digital media—in a sense it rejected them. It is not user generated, it is not social, it is not bite sized, it is not free." In almost every way besides "its route into people's homes," it is "the same as television . . . old-fashioned, passive, narrative entertainment."[37]

The recommender system's credibility—and thus Netflix's brand and ability to function as a business—depends on one first principle: the perceived infallibility of information collection and processing at scale. The effective suspension of this gravity-defying legend requires a perpetual game of hide-and-seek: like the Wizard of Oz, Netflix's data are ever-invoked, sometimes heard in impressive tones, and yet never seen. The recommender system is no standalone object. It constitutes Netflix's DNA and USP, the base to a larger superstructure: the mythology of big data.

A key part of the big-data myth, eagerly swallowed and spit back by the technology press and other news media, revolves around proprietary information processing. The company guards its data, including and especially viewership figures ("ratings" in the old parlance of broadcast television), much like Coca-Cola keeps its drink recipe secret: with coy affectation. Unlike television networks and film studios, whose ratings and box office are publicly available information, Netflix prevents even its content suppliers (major Hollywood studios, television networks, independent distributors) from knowing precisely who watches their productions, not to mention where, how often, and with which devices these viewers do so. Indeed, this imbalance of knowledge is Netflix's key point of leverage in acquisition negotiations, and a mystery in the public sphere.[38] With even film and series makers shut out from much if not most of their own data, the precision of Netflix statistics remains unverifiable, and thus subject to aggrandizing speculation.

Indeed, in tantalizing cases—such as the aforementioned academic article by Gomez-Uribe, the Amatriain and Basilico technology blogs, not to mention the Netflix Prize—the streaming company occasionally dangles a few numbers in ad hoc dumps. For instance, approximately once a month, Netflix releases a feel-good story with its proprietary data. About what shows subscribers binge and by which they "cheat" on their partner—in other words, watch episodes in advance of her or him. About which *Star Trek* episodes fans view most. About the series that users consume—the "gateways"—before going on to gorge themselves on popular superhero shows. About how teenagers feel closer to their parents when the latter watch favorite Netflix programs together with them. About mothers' efforts to squeeze in Netflix streaming while cooking, washing dishes, or performing other household chores. In short: the press releases feature as easy fodder for today's low-paid, cut-and-paste journalists. The last items were destined for parenting magazines, blogs, and popular online forums like *The Mommy Mix*, *Dad You Geek*, and *Mumsnet*, and helped boost Netflix's profile among early adult and middle-aged suburban women, a desired demographic after the company's initial advantage with geeky early-adopter urban men. Netflix's PR team reached out to these bloggers by inviting them to an "exclusive club," the "Stream Team," a means of influencer-based promotion for Netflix to "keep in touch with parents who love entertainment and enjoy blogging about it."[39] Partners received free subscriptions, merchandise, and tips on upcoming content in exchange for favorable posts that parroted talking points about the service and its recommender system. See, for example, this post from *What Katy Said*:

> Netflix challenged us to play a little game, to help us get used to our new account. We decided to get the girls involved and use the little fortune teller they sent to help us choose a family movie. . . . I think we may have a new craze in the house!
>
> Each of us have our own account, tailored to our own tastes. The girls have a kids account and it brings up suggestions based on what they have watched before. We logged into G's account for the challenge and wrote down the different categories, she then chose a movie per category and we wrote them under each flap. She was so excited to play the game and seeing as she is a little obsessed with phonics right now, she enjoyed sounding out the colour names.[40]

Netflix's fun-fact, listicle-ready press releases—which often come replete with poster-style, shareable graphics of statistics—pop up on Reddit and other aggregators and inform items in newspapers and blogs, providing Netflix with easy publicity in the form of coverage. The *New York Times* and other newspapers legitimate Netflix and its promotional rhetoric by reprinting the company's data-informed studies with the barest of critical frames. For example, one *New York Times* article claimed that 12 percent of Netflix users admit streaming in public restrooms—recycling wholesale a press release from three days earlier—and concludes, parroting the platform's usual digital revolution rhetoric, that "media consumption patterns and social customs are shifting." The piece makes just one critical claim: "Streaming companies, including Netflix, have been reluctant to share such data except when it serves their own interests."[41] The overall effect, however, is a gratefulness that Netflix deigned to enlighten us on the matter.

This master narrative is effectively rehearsed in the story surrounding Netflix's first major exclusive series, *House of Cards* (2013–2018); the founding tale of the Netflix "Original" resounds with the big-data myth. In 2013, the prodigious press coverage of *House of Cards* revolved not simply around its all-star cast, including Kevin Spacey and Robin Wright, or David Fincher's production and direction, or the adaptation of the BBC source material. The lede was Netflix's supposed use of its viewership information as a commissioning tool.

Press reports claimed that the series was the direct product of Netflix's collected data, a sort of algorithmic Frankenstein's monster, assembled by mapping the coordinates of millions of subscribers' taste data points about source material, visual style, genre, tone, director's vision, and acting talent. Dozens, if not hundreds, of high-profile commentaries—stoked by Netflix press releases and executives' public comments—described a futuristic vision of audiovisual storytelling. "In the television business," veteran *New York Times* media critic David Carr opined in 2013, "there is no such thing as a sure thing. You can have a gold-plated director, a bankable star and a popular concept and still, it's just a roll of the dice." With *House of Cards*, however, the "nobody knows anything" rule is no longer applicable; the "spooky part about" the series was that Netflix executives "knew it would be a hit before anyone shouted 'action.'"[42] Here Carr proposes nothing less than the conclusive revocation of films' and series' traditional status as risky experience goods; Netflix's data collection methods had made the success of an audiovisual narrative suddenly predictable. The sentiment behind his statement—that "big bets are now being informed by Big Data, and no one knows more about audiences than Netflix"—spread widely among the commentariat, in the tech, media, and business press, in broadsheets and television news. Scholars effectively reproduced this consensus in later academic accounts.[43]

Of course, Netflix itself cued and pushed this story, asserting a competitive advantage in almost science-fiction terms. Sarandos claimed in interviews that the series was "generated by algorithm."[44] Even *House of Cards* artistic staff who knew better perpetuated this backstory; Kevin Spacey, for instance, elaborated on this narrative in a speech to industry professionals.[45] According to these accounts, the future of film and series making was no longer the domain of the hit-or-miss gatekeeping establishment of out-of-touch Hollywood studio bosses and network executives. Finely tuned computers would effortlessly reverse-engineer tastes personalized to the whims, wishes, and wisdom of crowds.

The widely reported *House of Cards* production history, however, is a big-data illusion. My claim here is not to deny that *House of Cards* was a significant cultural intervention. To be sure, premiering a series with all of its episodes available immediately went against standard practices; major Hollywood talent like David Fincher, Kevin Spacey, and Robin Wright producing "television" was a rarity (although hardly unheard of) in 2013. By these more limited measures, the series certainly deserves notice for its innovation.

Nevertheless, all serious sources indicate a much different—and more normal—story of *House of Cards'* commissioning and production. Interviews with the heads of the production company MRC and other insider accounts reveal that the series arose in a manner not at all atypical in 2010s Hollywood. An intern at MRC recommended the 1990 BBC adaptation to cochairman and co-CEO Modi Wiczyk, who then, in June 2008, licensed the intellectual property from source material author Michael Dobbs and, in the summer of 2009, attached key talent, including executive producers David Fincher, Joshua Donen, and Eric Roth, who in turn hired head writer Beau Williamson in January 2010. Location scouts, budgets, storyboards, and production designs were commissioned; total development costs were placed in the high six figures. In March 2011—nearly three years after licensing the source content and after a year and a half of intense story development—the team took the show bible, the script of the first episode, and the names of attached actors, including Spacey, into network pitch meetings. The production team proposed the series to all of the major premium cable networks, including AMC, FX, HBO, Showtime, and Starz; at least two of them made a serious offer. MRC also contacted Netflix at this time but only with the intention to partner with the streaming service for a further window after the initial cable run. It was only when Netflix made a far superior bid—with much better financial terms (e.g., ordering two seasons rather than one and requiring no pilot), enhanced legal freedoms (including ownership of content), and promises of fewer artistic interventions—that MRC decided to partner with the streamer over the finer-pedigreed cable channels. Although credible accounts claim that Netflix presented MRC with viewer figures to suggest that the series would be a big hit with subscribers because of subject matter, cast, and crew, the VOD service hardly invented the series or seriously guided its aesthetic or thematic content on the basis of its data.[46]

In no way, then, did Netflix imagine or construct *House of Cards* via its collection of viewership data. At the most, one could argue that this information served to vet the viability of the project; Netflix's numbers *backed up* its very human decision to acquire the series. We can perhaps best characterize the "innovations" in the acquisitions arrangement as concessions made by a company desperate for content and with comparatively little leverage. After all, at the time of the deal, Netflix had never distributed its own series, did not have publicly available ratings, and had little indication of the marketing muscle of an HBO and Showtime, risks that threatened to eradicate the show's value internationally and in subsequent windows. In 2013, however, the mythmaking stuck that Netflix could be unassailable because its vast treasure trove of proprietary data would make every acquisitions

FIGURE 3. Symptoms of contemporary pop-culture beliefs, *Moneyball* (2011)—pictured here—*Minority Report* (film, 2002; series, 2015), and *Money Monster* (2016) explore the dreams and nightmares of algorithms and data collection. Credit: AF archive/Alamy Stock Photo.

decision—one of the most fraught in the notoriously risky entertainment industry—essentially a no brainer. Perhaps we can understand such views better alongside other contemporary pop-culture beliefs about algorithms and data collection. In those days, quantitative data analytics featured in a series of films: what in *Minority Report* (2002) was still science fiction found realist expression in *Moneyball* (2011, Figure 3) and then satire in *Money Monster* (2016). Despite the difference in genre (science fiction, sport drama, thriller/black comedy) and subject matter (criminal justice, baseball, finance), each of these films explores the supposedly magical social possibilities—or entertains the dystopian personal ends—of big-data dreams. In the context of their respective productions and initial receptions, each seemed to reflect and further fuel popular (and no doubt industry) beliefs that hunches and hedges had no place in modern business: follow the data became the new motto.

Ultimately, the big-data mythmaking that undergirds the recommender system, and Netflix as a company and cultural phenomenon, is a means of *de datis non disputandem est* control: "you can't argue with data" becomes a mantra-like means of foreclosing critical discussions with the most positivist form of empiricism. This is particularly evident in a widely reported feature of the Netflix recommender system (one that is integral to the arguments of those, such as Smith and Telang, who claim that conventional entertainment executives do not trust data): the algorithms supporting the recommender systems supposedly offer patterns and knowledge unavailable to and indecipherable by the human mind, in particular because of cultural biases. Harping on the recommender system's black-box, unfathomable internal logic, commentators speak in extraterrestrial terms: "There's a sort of unsettling, alien quality to their computers' results," the *New York Times* ventured about Netflix. "Possibly the algorithms are finding connections so deep and subconscious that customers themselves wouldn't even recognize them."[47] The Netflix "algorithms analyzed the patterns" derived from viewers' preferences, Keating reports, "and assigned its own descriptors to films that were richer and more subtle than labels like director, actor, and genre but had no real meaning in the human mind."[48] Netflix "collects and dissects" data in its recommender system that supposedly reveal "surprising correlations: For example, viewers who like *House of Cards* also often like the FX comedy *It's Always Sunny in Philadelphia*" (2005–present).[49]

Alexis C. Madrigal's *Atlantic* piece is paradigmatic of this stubborn trend in Netflix reportage. After self-congratulation for reverse-engineering Netflix's microgenre categories (which nevertheless, in his descriptions, seemed to take him not much time at all using mostly publicly available freeware), Madrigal concludes on a mystical note, surprised at the popularity of *Perry Mason* (1957–1966) episodes. "There is *so much more data* and a whole lot more intelligence baked into the system than we've captured," Madrigal writes. "It's inexplicable with human logic. It's just something that happened." He continues: "The vexing, remarkable conclusion is that when companies combine human intelligence with machine intelligence, some things happen that we cannot understand."[50]

If anything, these observations symptomatically confirm that coders (and *Wired* editors and *Atlantic* feature writers) remain susceptible to human biases. In order to be surprised that Netflix's most preferred microgenres contain *Perry Mason*, one would have to expect that in 2014 *Perry Mason*—a long-running series in perpetual syndication—would be significantly less popular than *Lord of the Rings* or Robert DeNiro. That is, one would have to be naive, possessing limited knowledge of cultural and industrial history. Spinning a mythology of "only-the-computer-knows" microgenres and "surprising correlations" belies the fact that such connections are legible to any film or media scholar. Yes, on the bluntest level of genre *House of Cards* is a political thriller set in the White House and *It's Always Sunny in Philadelphia* is a sitcom that takes place in a bar; the former has cinematic production values and A-list Hollywood talent, while the latter is harshly lit and debuted with a cast of unknowns. And yet both series are dialogue heavy. Both contain a similarly cynical worldview, narcissistic characters, transactional relationships, and a high degree of moral relativity. Does it take an algorithm, let alone a PhD, to uncover such "mystifying" or "surprising" correlations? Netflix and its cheerleaders in the tech press have reverse-engineered insights available to a second-year undergraduate in film and media studies.

Indeed, one wonders, if the algorithm is so predictive, why did it "generate" *Gypsy* (2017), *The Get Down* (2016–2017), *Marco Polo* (2014–2016), *The Letter for the King* (2020), *The English Game* (2020), *October Faction* (2020), or any of the many low-quality, poorly reviewed, and presumably little-watched series that Netflix commissioned and then canceled after their first or second season?[51] Of course, the company might publicly claim that their computers are not the problem, but rather content makers' slipshod execution. But these are always the excuses of big-data ideologues. Netflix's Wizard-of-Oz brand of public relations—by which it rouses great expectations with data that the user is expected to simply accept sight unseen—hardly inspires confidence. Does it take a machine to determine that viewers in 2017 might like Brad Pitt and Will Smith movies or in 2020 a film starring Anne Hathaway, Ben Affleck, Rosie Perez, and Willem Dafoe and based on a Joan Didion book? This was the conceit of the supposedly daring gambits of *War Machine* (2017), *Bright* (2017), and *The Last Thing He Wanted* (2020). In any event, the finished products suffered quick deaths, algorithm or not. Pitt's $60 million dud, "Netflix's attempt to offer the equivalent of mid-budget, star-driven and slightly prestigious Hollywood fare that the major studios have mostly stopped making, is a fascinating miss," according to one critic.[52] Reinventing the wheel, Netflix seems at least as prone to experience-good, nobody-knows failure as any other media institution, data be damned.

PRODUCT DIFFERENTIATION AND HIGH-TECH, LOW-SPEC RECOMMENDERS

Why does Netflix cue, cultivate, and reinforce the myth of big data? Before returning to engage with Netflix's critics on how the recommender system is supposedly replacing humanistic cultural mediation, I must say a few words about why the company expends so much effort on the "performed backstage" and recommender-system mythmaking.

The credibility of big data and algorithms is akin to that of critics. Recommender systems need to build, elicit, and cultivate trust as much as Peter Bradshaw, Manohla Dargis, "the *New York Times* critic," or an institution such as *Sight and Sound* does. Although the basis of the authority to suggest differs between recommendation forms, credibility must be performed and rehearsed regularly in order to gain or maintain the power to persuade. Ultimately—and unlike some of the most well-known and respected critics—Netflix maintains fundamental weaknesses as a business proposition, above all because of strong competition and the ability to replicate the company's model. This creates the imperative to alchemize trust by other means: the myth of big data.

As we have seen, the recommender system serves important economic imperatives internal to the company. In different ways, both in the DVD days and now in the almost exclusively streaming era, the system has helped focalize viewers' attention in the face of content overload and maximize the use of the catalog's breadth. Netflix claims that the recommender system has a statistically significant advantage over simply suggesting titles of overall popularity; its algorithms create a supposedly billion-dollar premium in terms of retained subscriptions and decreased spending on attracting new customers, set-up costs, free trials, and so on.[53]

The rhetorical attention to the recommender system's personalization, effectiveness, objectivity, and other big-data mythmaking, however, gestures to a perhaps more important symbolic function. It helps differentiate the brand from other vehicles of motion picture entertainment, from Amazon to HBO. In this context, it is important to note how and why the importance of the recommender system in marketing messages has increased—and not only because Cinematch has become more sophisticated. In the early years, as a US-wide DVD-by-post service, Netflix's chief competitor was the video rental store (and especially the big chains like Blockbuster) and its USP was clear: a much wider selection, the ability to order online, and a flat-rate subscription price without late fees. Nevertheless, as the company expanded internationally, it has (except for a few million subscribers in the United States) abandoned DVDs for streaming, significantly shifted focus from feature films to series, and moved considerably toward a medium-term catalog goal of half licensed content and half commissioned productions. This reinvention catalyzed profound changes in the company's business plan and indeed the corporate sector as a whole. Once video stores, Netflix's competition now encompasses all manners of content providers, aggregators, distributors, and producers including Amazon, YouTube, Disney, Apple, HBO, Sky, Comcast, and DirecTV. With the company's stated objective, at least since 2012, "to become HBO faster than HBO can become us," Netflix has above all stressed its "personalization"—its recommender system and data collection—as its prime selling point in this competition. According to executive Sarandos, services such as HBO Now "will start to attack us on the things we believe we do better than anyone else. Subscription. Personalization. Encoding. Multiplatform delivery. We need to differentiate ourselves on all fronts. Our data and algorithms help us perfect personalization.... Ultimately we want to produce our own shows to have control."[54]

Pursuing this sort of differentiation bespeaks, on the one hand, the relatively long development and relative (input) sophistication of Netflix's system. On the other hand, however, it responds to the nakedness of the company's offer compared to other providers: Netflix sells one product. In contrast, Amazon offers an array of commercially popular audiovisual content essentially as a fringe benefit to receiving free, speedy shipping (same-day, overnight, or two-day, depending on the country and city). Other companies and potential competitors have used various synergies and gimmicks to add value to the VOD proposition.[55] Most recently, Disney and Apple, launching their subscription streaming services Disney+ and Apple TV+ respectively, have taken on Netflix with a lower price point, hardware tie-ins, or special access. These moves expose Netflix's vulnerability and also deplete some of its more popular offerings: Disney now offers its and Pixar's content exclusively on its own platform. Even though Netflix now has a production output that exceeds both the largest American television networks and Hollywood movie studios, compared to Disney it still lacks an extensive decades-long back catalog of recognized content.[56] This weakness became even more pronounced in 2019, when Disney acquired 20th Century Fox, including the latter's prodigious library of film titles.[57] In sum, Netflix faces serious business risks, including the replicability of its model, low barriers to enter the VOD market, behemoth conglomerates leveraging their catalogs or cable wires, and so on.[58] Within five months of launching, Disney+ had already attracted fifty million customers, nearly a third of Netflix's total subscribers after twenty-three years as a going concern. By early 2021, it had surpassed 100 million and industry analysts were forecasting that, at the present rate of growth, Disney would overtake Netflix as the world's most-subscribed streaming provider in 2024.[59]

Given this increasing competition, how can Netflix differentiate itself and claim added value? With catalog acquisition pricing for VOD rising significantly because of proliferating bidders, Netflix can only double down on its personalization algorithm and expand its range of original content. The company is betting on the entire entertainment film and television industry shifting toward its on-demand, subscription model. With most consumers unlikely to pay for more than a handful of such services, Netflix seeks a significant comparative advantage with a considerable head start on recommendation technology and data collection methods. Already its corporate investors rate and value the company as a matchmaker—that is, for its recommender function of linking consumers to products—rather than as a content provider, an impression only buffeted by its distinct flexibility about what it streams in terms of format (film, series), genre, aesthetic, or virtually any other conventional parameter of the moving image industry. *Pace* Stuever, not having an identifiable aesthetic identity can be advantageous. Indeed, for the Wall Street types who fund the company and its frequent debt-market asks, Netflix has more in common with Uber or Match.com than with Disney or HBO: the health and competitiveness of its suggestion algorithms—and their feature as a key marketing and retention tool—represent the be-all and end-all of the company.[60] What these new-media investors neglect to recognize: this system also replicates critics' traditional matchmaking function, guiding and shaping which content audiences see.

Netflix's success, which depends so heavily on consumers' belief in the recommender system's credibility plus a steady stream of appealing original productions, is far from assured. Despite the fawning press reports of "global dominance," balance sheets are by all accounts precarious, with huge headlines spends—each year increasing by ten-digit increments—masking similarly prodigious net liabilities.[61] After disastrous reactions to a 2011 plan to devolve its DVD service and raise prices, Netflix once again raised prices in 2017, 2019, and 2021, with a tiered structure for better-quality streaming and increased access across more devices. These actions underline a basic reality: finances are still floating on investment capital, rather than a truly sustaining subscription base, even at over 200 million members worldwide. According to reliable reports, the company is hemorrhaging $3–4 billion per year.[62] Despite its marketing narrative around consumer-friendly personalization, Netflix has committed a series of unforced service-related errors that have led to customer-relations debacles; it continues to draw bad press for creepy data dumps and a shrinking portfolio of content.[63] From the perspective of business survival, Netflix needs Cinematch and, in particular, the big-data mythology surrounding it to drown out the noise. Ultimately, everything else the platform offers can be done by competitors with more attractive benefits.

This, then, is the central paradox: despite the fundamental weaknesses of Netflix and its recommender system, there is no dearth of executives' public bravado, nor end to the corresponding reams of fawning media coverage. In equal and opposite measure, scholars have estimated Netflix and its effects on audiovisual media selection, tastemaking, and criticism as unprecedented and profound. By way of conclusion, let us reengage with some of the most important critiques, which ascribe Netflix's recommender system with culture-altering effects on taste and cultural authority.

Alexander, Arnold, Smith-Rowsey, and Striphas and Hallinan represent some of the smartest commentators on this issue. Although I share their concerns, their conclusions remain premature. As previewed in the beginning of this chapter, the critique can be distilled into a few overlapping theses: a new "posthuman" form of cultural arbitration is emerging; its basis, data collection, relies on reductive quantitative measurements that are concealed from the public; data-driven engineering has usurped the cultural authority from qualitative humanistic evaluation (i.e., critical reviews). We need to reassess these ideas for the ways in which they may unwittingly overestimate and mystify the recommender system's use, capabilities, and credibility.

Despite the "posthuman" rhetoric—a curious echo of big-data myths— the Netflix recommender system is not divorced from human input and design. Scholars have raised many legitimate objections regarding humans' lack of agency in the Netflix recommender system, highlighting for instance the fact that users cannot create a microgenre or otherwise actively guide their recommender system experience. And yet the system functions by a process in which humans have created the basic categories and labels, the essential semantics and syntax. Humans apply tags to the underlying content, using their interpretative judgment to assign a numerical value to the level of romance, violence, and so on. Humans license or commission the films and series themselves; they compose the thumbnails by which the content will be represented on screen. Humans design system algorithms and thus chart the pathways of machine learning. In this sense, Netflix recommendations— despite all the company propaganda and fawning coverage to the contrary— are neither entirely computer-driven nor truly "objective." Less controlled by artificial intelligence, these systems remain guided by individual and collective human cognition, logic, and decision-making. Indeed, as Cathy O'Neil and Safiya Umoja Noble have pointed out, such systems contain all of the cultural stereotypes, racism, and sexism that one would expect from a human-generated enterprise.[64]

Humans dictate the operations that computers execute; in and of themselves, algorithms constitute on-off switches, reducible to the commands *and*, *or*, and *not*, that their human masters set up and arrange.[65] Algorithms cannot *produce*, but may certainly *reproduce* biases, unconscious or not, that already exist in their developers' culture. Indeed, we have seen the ways in which programmers have designed recommender systems for diversity—at least when executives believe this will lead to more "engagement," and thus profits. To be sure, objections to Netflix's microgenres are well founded. As one commentator notes, categories like "African-American Movies" or "Series with a Strong Female Lead" remain "culturally coded through 'difference'" to an implied normative white male identity.[66] And yet such classifications reveal the (conscious or unconscious) biases of the company's employees and their American cultural context, not of the algorithms themselves. "To accuse an algorithm of bias," Tarleton Gillespie argues, "implies that there exists an unbiased judgment of relevance available, to which the tool is failing to hew."[67] Even Netflix—as a corporate enterprise economically invested in, and thus ideologically committed to, the infallibility of data and data-derived insights into taste—openly admits a "seventy-thirty mix" between data and human intuition, respectively, "and the thirty needs to be on top."[68] Elsewhere, Sarandos submitted that "big data is a very important resource to allow us to see how much to invest in a project but we don't try to reverse-engineer."[69] Chief product officer Neil Hunt clarified that "that's not to say that statistics will always impact a storyline. For example, if a director is looking for a supporting actress, Netflix can query its data to come up with the 4 actresses that best resonate with the target audience for the show they're producing. But if the director decides that the actress, despite being popular, has no chemistry with the lead, he or she can veto that suggestion."[70] Recognizing algorithmic recommender systems as yet another human-designed (if not human-executed) form of cultural suggestion attenuates the force of both Netflix proponents' and critics' arguments.

The second major criticism revolves around the concealed nature of the recommender system and its quantitative data collection processes, the impenetrability of algorithmic systems and their data sets to most humanists and social scientists: the so-called black-box problem. Boltzmann machines, Markov chains, Bayesian networks, support vector machines, matrix factorization: How can one address a problem, these commentators ask, whose underlying details are neither visible nor accessible, subject to constant change, and composed in a language most humanists lack the training to understand?[71] If we could only "see" the data and formulas, these commentators insist, we could begin a productive discussion. Because "such algorithms are at best invisible" and their "criteria, assumptions and forms of classifications are inaccessible to and beyond the control of the individual consumer," one scholar claims, such devices have untold and likely devious effects on taste, not to mention pernicious skews "to the provider's commercial or political benefit."[72] Indeed, according to one commentator, recommender systems' "influence is now so deeply embedded and often unseen that it is increasingly hard to imagine or describe the impact that algorithms might be having on cultural encounters, tastes, preferences and subsequently on communities, groups, networks and movements. They are already too numerous to describe or list."[73]

Nevertheless, despite some scholars' nothing-to-be-done tone, these issues seem neither insurmountable nor unprecedented: cultural problems do not always come in forms suited to academics' ease of use. Media companies throughout history have attempted to hide, disguise, or dress up their working methods, decisions, and sales statistics. For example, historians have shown how advertisers and product designers moved in lock step in the 1920s to consciously mystify the linguistic formulations and the product design behind many household goods. Whereas Ford's Model T could be repaired by any owner without formal training, new brands and types of equipment and appliances aimed for a complexity that would overwhelm the consumer. Jargon like "halitosis" brokered new demand for hitherto unimagined problems in deliberately technical advertisements.[74] These procedures and the publicity surrounding them stoked an attractive aura of scientific quality—advertising equated a sophisticated new form with a vastly improved use outcome—at the same time that they created new markets for repairs and updates. Planned obsolescence replaced users repairing products themselves. Of course, these machinations resemble the myths that Netflix and Amazon peddle, with their invocations of PhD-level math and promises of revolutionary improvements in consumer satisfaction. The black-box discourse veils, in new-media language, the quotidian difficulty of access that any critical media industries scholar faces and must endeavor to resolve creatively. These supposedly intractable methodological conundrums are not incomparable to securing forthright interviews with aloof industry executives, who often have little incentive to reveal themselves to academic scrutiny.[75] Such arguments assume that for-profit media companies try to "understand" their customers differently than other companies do; ultimately, however, all use market data, whether emanating from a cookie or from a Nielsen logbook.

To be sure, it is essential that we scrutinize businesses such as Netflix and Amazon and their nontransparency in collecting and using personal consumption data. And yet many critics seem to imply that taste was never quantified before; that we are heading for a dystopia whereby taste will *only* be quantified; and that algorithmic recommendations are unavoidable, inescapable, hidden, all-knowing, culture-shifting, and fundamentally new.

In fact, there is a long history of attempts to quantify taste, particularly in test screenings, studio focus groups, television ratings, cinema box office, and myriad industry reports. Media companies have long gauged viewer engagements using numerical figures, often classifying consumers by age, gender, income, education, class, or race in order to fine-tune marketing and even modify the content itself (narrative resolutions being a traditional variable). One wonders whether Standard Oil or the old AT&T or MGM cared about specific people or rather saw them as interchangeable customers, important only to the extent to which they belonged to demographic quadrants, essentialized only to age, gender, race, geographic location, and, in general, the extent to which they might generate revenue or not. In the words of Evgeny Morozov, challenging the exceptionalism of algorithmic computer processing and big-data internet companies, "Nothing has changed about the epistemic assumptions about how capitalism operates that made companies suddenly realize that data was valuable. It was always valuable but very hard to grab before."[76]

In turn, for better or worse, the entertainment business has long fudged and falsified audience data, hidden the information as proprietary, or weaponized it in PR stunts.[77] Other forms of recommendation and other means of optimizing cultural products (focus groups, market research, ratings) have not been wholly transparent to users and commentators. The film and television industry and its means of production and marketing have always aspired to disguise inner workings, the bare ceilings and columns of the "dream factory"; Netflix is hardly unique in this respect. Although naturally one wishes that the company and its competitors were more forthcoming about why a certain title is recommended, this concealment of data represents merely an *attempted* elitism, the attempt to produce an asymmetry of information. One need not be an expert in matrix factorization, nor does one need to investigate the trade secrets of every last algorithm in Cinematch to understand the basic contours of its system. As Bernhard Rieder has demonstrated, although "algorithms used in concrete systems may often be hard to scrutinize, they draw on widely available software modules and well-documented principles that make them amenable to humanistic analysis."[78] We must not default to black-box fallacies: the belief that that which I cannot see must be dangerous. Nor should we assume that greater transparency would somehow detoxify or enervate corporations' profit motives. Objectors' arguments unwittingly parallel those of business oracles who preach the direct relationship between more data and more satisfied customers, more profits, and thus better outcomes. Let us recall, from previous eras, similar objections to nontransparency regarding what are now legacy recommendation forms. In a 2007 tome on criticism, ratings, and reviewing practices, for example, Grant Blank wrote that the "process in a connoisseurial [i.e., critic's] review is a black box. Readers cannot easily see how evaluations were reached." Such reviews remain thus "prone to ethical problems."[79]

Yes, the volume, variety, and velocity of the data and data-acquisition methods at Netflix's disposal may be pioneering. The transition from the representative sampling of audiences toward universal data collection in real time is likewise unprecedented. Nevertheless, the big-data myths rehearsed in the *House of Cards* production history and myriad related press releases about personalization of content belie the fact that, at best, aggregated audience groupings (rather than individual spectators) *influence* commissioning decisions. Despite the granular data Netflix may have (starts, stops, pauses, scroll-overs) and the considerable anxiety this creates among cultural commentators, what is the actual value of knowing that I paused *Stranger Things* (2016–present) series 3 episode 3 at minute 23? At the individual level at least, precisely zero.

We live in an age in which, by many measures, data sources are as readily available as they ever have been. In this context, labeling potential increases in hidden or nontransparent data as "posthuman" or "socio-technical assemblages" itself resembles an operation of science-fiction obfuscation, a mystification of culture and taste. For this line of reasoning, taken at face value, suggests that preferences are unique and ultimately unknowable: *de gustibus non disputandem est* presented as an intellectual program. Ideological conviction, rather than proven fact, seems to drive the claim that data and mathematics cannot help us approximate features of taste in any way;

I respectfully disagree with the implication that quantifying taste is a priori a reductive proposition. For this obscures cultural preference under a romantic veil—something alien, fundamentally innate, hopelessly complex, and ultimately indecipherable—that curiously resembles Netflix's and competitors' mystification of big data and algorithms, a Wizard of Oz.

Faced with the technical intricacy and operational opacity of Netflix's recommender system and persistent marketing rhetoric, commentators fear the worst and ascribe the tool with a *cultural* complexity. They equate sophistication of the input for usefulness (or deviousness) of the suggestion output. Many see Netflix like James Bond or the NSA: an operator capable of powerful and clandestine surveillance, a Big Brother. The company reinforces these perceptions with incessant aggrandizing big-data mythology and cues PR-fed news reports of a "new world order" of media.[80] But is Netflix, or any other VOD provider for that matter, creating something so genuinely novel and widely used that it is shifting culture in such a fundamental and permanent way?

In order to resolve this query, we urgently need to evaluate, systematically and rigorously, the credibility and effectiveness of these systems as experienced by *real users*. Fantasies of algorithms solving cultural quandaries of recommendation—and thus initiating a new phase of business innovation and profit making—remain pervasive among many commentators. Nevertheless, in practice, many of these services deliver more dubious results. The key user complaints about VOD services with personalized, algorithm-based recommender systems revolve around the inadequacy of recommendations and the lack of selection.[81] In internal memos and closed-shop reflections, Netflix engineers acknowledge the many "open problems" they must still solve within the system, which remains a work in progress.[82] In semi-private settings, even the company's industry-facing representatives admit that the recommender system is hardly flawless. Personalization of content, Cees van Koppen, Netflix's manager of public policy for Europe, the Middle East, and Africa, told a group of industry insiders and government officials at a 2017 London conference, remains "one of the biggest challenges." He continued: "We believe we are in the early stages," confessing, "I still sometimes get recommendations that maybe are not entirely correct."[83] Netflix may claim that a certain film or series has a 99 percent chance of engaging its user. The reality, in any given moment of decision-making, is far more complex.

To be sure, there is a "whole lot" of sophisticated—yes, PhD-level— mathematics and programming involved in the Netflix recommender system. The Netflix Prize's first-year leading team needed over two thousand hours of work and 107 algorithms to achieve an 8.43 percent improvement on the existing system. The winning team blended "hundreds of predictive models" to reach the goal of improving Cinematch by 10 percent.[84] Nevertheless, I submit, some commentators conflate mathematical complexity with effectiveness; they confuse automation and pervasiveness with useability and usefulness. From many, if not most, viewers' perspectives, sophistication of input is more or less trivia; only nuance of recommendation constitutes true innovation. A user-centric perspective can enrich, complement, and correct analyses of technology and form alone, generating more differentiated assessments of cultural effects. The following chapter embarks precisely on such an empirical audience study.

FIVE

How Real People Choose Films and Series

THE RHETORIC SURROUNDING RECOMMENDER SYSTEMS has largely polarized into two divergent narratives. In succinct terms, the first highlights consumers' supposedly unprecedented access and choice. The second posits malevolent technologies and profit motives guiding ever-narrowing tastes. Both stories, curiously, stipulate highly effective, powerful, and historically unprecedented tools. And both sides of the debate tend to make claims about uses and effects without engaging with actual users of recommender systems, preferring to infer consumer behaviors and social consequences on the basis of technological forms: in other words, technological determinism. To be sure, a few critical user studies of these systems exist: most of these, however, are small in scale, not squarely on point, or were conducted while such applications remained in their infancy.[1]

This chapter aims to redress these lacunae by making real users and uses of recommender systems its central focus and direct object of inquiry. It dwells on four key research questions: (1) To what extent are recommender systems actually used as a primary means to select and access films and series? (2) To what extent do users understand the mechanics of algorithmic recommender systems? (3) What level of trust, credibility, and effectiveness do users assign recommender systems vis-à-vis other sources of information and suggestion? (4) In what ways do users talk about their use of recommender systems in the context of choosing and experiencing films and series?

The appendix provides a fuller explanation of the research design, including the question-and-answer sets as well as the interviewees' basic demographic information. By way of summary, the raw data emerged from two central procedures. The first was the quantitative analysis of two nationally representative surveys, which I designed and YouGov Plc carried out online.

Total sample sizes were 2,123 UK adults and 1,300 US adults. The fieldwork was undertaken 13–14 November 2018 (UK) and 13–15 November 2018 (US). The figures have been weighted and are representative of all UK and US adults (aged eighteen and over), respectively.

The second procedure entailed the qualitative analysis of thirty-four in-depth interviews. The twenty female and fourteen male participants, living in six different regions of the United Kingdom, ranged in age between their teens to their sixties. The sessions, each of which lasted roughly twenty-five to fifty minutes, consisted of a semi-structured interview plus (in most cases) a demonstration: participants logged into their favorite VOD portal (most often Netflix) and showed the interviewer how they go about choosing content. Because of the small sample size—and squarely in the tradition of qualitative audience research—the goal of the interviews was not to come to generalizable claims. Rather, the interviews aimed to triangulate, deepen, explore, and explain my national surveys (and polls commissioned and undertaken by other research institutions), to illustrate real ways of speaking, to probe deeply, and thus to enrich the blunt quantitative snapshots.

Taken as a whole, the mixed-method audience study reveals a diversity of ways that real users engage with and understand recommender systems in the context of choosing and watching films and series. Compared with industry players, business journalists, media critics, and university academics, ordinary laypeople deploy their own unique idioms to speak about Netflix, VOD, recommender systems, and their audiovisual content choice. These ways of speaking constitute folk understandings, which often include a mixture of factually correct and incorrect assumptions, some deriving from and transposed via journalistic and publicity discourses. Some readers might find the interviewees' responses at times naive, misguided, or paranoid. Nonetheless, they reflect how real users talk about, and understand, their selection and consumption of films and series.

Benjamin Toff and Rasmus Kleis Nielsen define folk theories as "culturally available symbolic resources that people use to make sense of their own media and information practices."[2] In this chapter I use this theoretical approach, and the general parameters and procedures of mixed-methods empirical audience research, to analyze how *real*—not imagined or inferred—users come to terms with their own film and series consumption in the age of Netflix. By way of preview, the evidence presented in this chapter reveals the following answers to the headline research questions formulated above: (1) Although SVOD use has become widespread in the United Kingdom and United States, recommender systems feature as the primary reference point for only a small subset of users in a small subset of viewing situations. Even those who say that they are likely to use recommender systems tend to speak about the tools within an iterative, multistage process of decision-making. (2) There is little evidence to suggest that lay users have an advanced technical knowledge of algorithmic recommender systems from a computer science standpoint. However, they exhibit an adequate functional understanding of the systems, most likely derived from and informed by company publicity, news media reports, and their lived experience of these and cognate (Google, Spotify) systems. (3) Although there are clearly small minorities of users who have very high or very low confidence in algorithmic recommender systems, ambivalent reckonings were much more typical. Users deployed a number of overlapping folk theories such as (a) "there is something cynical in me that mistrusts them . . . but as a consumer it's important" (rationalizing consumerist convenience with theoretical mistrust); (b) "it doesn't know how I'm feeling that evening" (algorithms mistake moods or remain inadequate to understanding human taste); and (c) attitudes on questions of variety and filter bubbles that imply resignation to fate ("there is no way you can run away from [algorithms]") or expressions of human agency and free will ("you can just go to the different genre"). (4) Inductively analyzing respondents' ways of speaking about these topics reveals several identifiable user typologies (e.g., Information Limiters, Information Gluttons, Lazy Choosers, WOM Providers) and themes (e.g., Keeping Up). These suggest that amounts of media use—and crucially, the importance of media use in their lives—figure as key components of the folk understandings and theories that people construct about how they choose content and negotiate VOD recommender systems.

I. TO WHAT EXTENT ARE VOD RECOMMENDER SYSTEMS ACTUALLY USED?

The quantitative results of the survey revealed that as of late 2018, among a representative sample, only 19 percent of UK adult residents (see Figure 4) and 20 percent of US adult residents (Figure 5) reported never having used a VOD streaming service. It also indicated that most of those 80–81 percent who said they had viewed films or series via these means were users of platforms (Netflix—UK: 51%; US: 63%; Amazon Prime Video—UK: 33%; US: 40%; BBC iPlayer—UK: 59%) that use algorithms to suggest content.

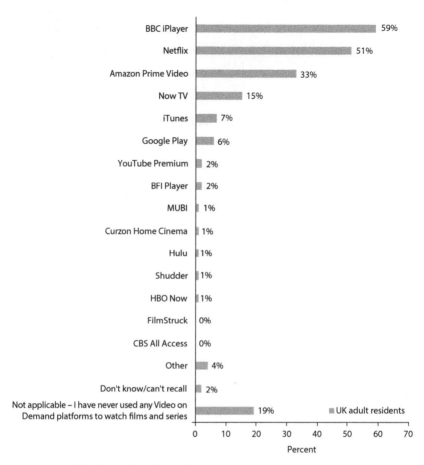

FIGURE 4. VOD Service Use, United Kingdom

Which, if any, of the following Video on Demand platforms have you ever used to watch films or series? (Please select all that apply. If you have never used any Video on Demand platforms, please select the "not applicable" option.)

Unweighted base: All UK adults online (*n* = 2,123)

Indeed, these were the most frequently used VOD services, in contrast to curation-style services (e.g., MUBI—UK: 1%; US: 1%). However, such quantitative insights—as powerful as they are—do not take into account whether algorithmic recommender systems remain the only or even primary means to select audiovisual content. It does not follow, in other words, that 80–81 percent of UK or US residents are blindly following recommender systems to make their film and series choices. Accessing Netflix to watch *The Society* (2019–present) on account of a billboard or friend's tip is just one

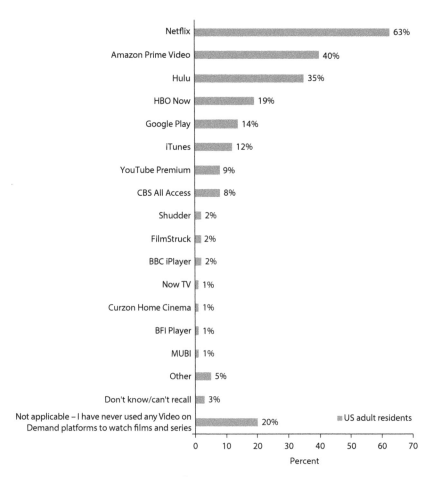

FIGURE 5. VOD Service Use, United States

Which, if any, of the following Video on Demand platforms have you ever used to watch films or series? (Please select all that apply. If you have never used any Video on Demand platforms, please select the "not applicable" option.)

Unweighted base: All US adults online ($n = 1,300$)

commonplace example of how VOD platforms accommodate traditional forms of suggestion beyond the in-house recommender system alone.

Even discounting the decisive fact that cinema, DVD, and linear broadcast television still constitute significant channels of consumption—where algorithmic recommender systems may play at most a secondary role, for instance, seeing a trailer recommended on YouTube and then deciding to go watch the film in the cinema—the survey results imply that US and

UK residents are still using a variety of traditional sources. For example, when (UK) survey participants who had used VOD were asked specifically about watching films and series on VOD platforms (as opposed to cinema, television, DVD, and other channels of dissemination), word of mouth (51%) still far outpaced genre search (24%), trailers on the platform (24%), descriptions of the content on the platform (20%), popular (15%) on or newly added (14%) to the platform, on-screen personalized recommendations (13%), and prominence on the VOD home screen (7%) as the source most likely to guide their choices (Figure 6). Furthermore, the opinion of family and friends while watching VOD as a group (26%), critics' reviews (19%), articles about a film or series or interviews with the actors or director (15%), user ratings (12%), review aggregators such as Rotten Tomatoes (8%), and user postings on film sites or blogs (8%) were other external sources that many viewers routinely sought out while making decisions about VOD content. When asked to *rank* their top sources overall, only 10 percent indicated "content recommended on a Video on Demand platform (e.g., Netflix, Amazon Prime) based on my prior viewing history" as among their top three.[3] VOD recommender systems, in other words, featured among UK and US adults' *least used* and *least trusted* sources for film and series recommendations. Indeed, substantial majorities said that, if forced to choose, they would be more likely to trust human critics (UK, 74%; US, 64%) over computer algorithms (UK, 7%; US, 12%) to provide a better film or series suggestion. These initial results already put into serious doubt the thesis that legions of viewers blindly trust Netflix—and only Netflix—to guide their consumption.

The interviews provided richer insights into how these functions work in practice. To be sure, several interviewees confirmed the fact, suggested in the surveys, that different viewing situations lead viewers to deploy somewhat different information regimes.[4] In other words, users are bound to use different sources and information-seeking routines when planning a trip to the cinema versus an evening with Netflix, but also depending on whether they are watching alone or in a group, actively choosing for a special occasion or selecting for a distraction at bedtime or while cooking or eating. Generally, the interview participants deployed a more intense and also more varied procedure for soliciting information before a cinema viewing. Participant 33, who regularly shows up at his favorite local cinema and chooses among the films on offer without even consulting an online schedule beforehand for the "adventure" is an exceptional case that proves the rule: the higher

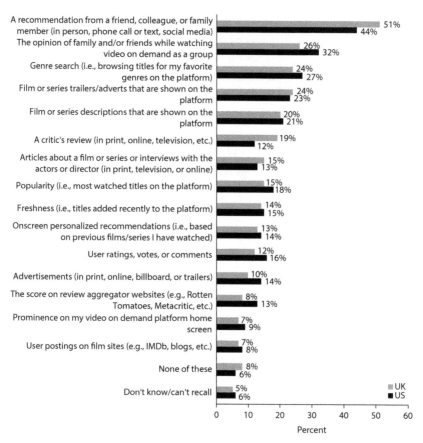

FIGURE 6. You previously said that you have used at least one Video on Demand platform to watch films or series. Thinking about when you have chosen a film or series to watch on a VIDEO ON DEMAND platform . . . Which, if any, of the following are MOST LIKELY to influence your choice of what to watch? (Please select all that apply.)

Unweighted base (for UK survey): All UK adults online who have used VOD platforms to watch films or series ($n = 1,692$).

Unweighted base (for US survey): All US adults online who have used VOD platforms to watch films or series ($n = 1,042$).

opportunity costs of cinema viewing versus SVOD mean that most all of the participants performed more research when planning to watch a film in the cinema. Most interviewees reported undertaking less research to prepare for VOD viewing, with some suggesting that much of it (besides pre-knowledge created by word of mouth and advertising) was mainly internal (image and description and trailer and sometimes match factor) to Netflix itself.

Indeed, among a number of the study participants, there was some evidence of what one might call a *one-stop-shop* phenomenon. That is, these participants (e.g., #4, #5, #16, #19, #20, and #22) noted that their use of streaming services, and in particular sophisticated platforms such as Netflix, may have led to consolidation of the variety and location of sources by which they select content. Participant 22 noted how when she wants to go to the cinema, she consults a wide variety of sources beforehand, whereas for Netflix "I just go directly to the description." In the words of Participant 20, "Usually when I open Netflix I've got loads of time, so I don't really care about how long it's going to take. And then I usually watch the trailer; I didn't used to do that when it opened, but . . . I used to find it very tedious, then you had to google or go on YouTube and try to find a trailer." According to Participant 16, "No, I think, on Netflix, I will watch something really easy-going, like really easy, and in the movies, I will like something—I don't know—longer and intense, I guess." The behavior of these participants suggests that, for some viewers at least, the convenience of Netflix's interface (where image, synopsis, cast and crew information, and trailer are built in) means they may be less likely to consult, for instance, critics' reviews or other sources beyond the platform.[5]

All of these responses, however, couple one-stop-shop comments with a sense that VOD is fundamentally and qualitatively different to other viewing experiences: note how Participant 20 suggests that when she watches Netflix she always "has loads of time" and is watching alone.[6] Early in the interview she reported her favorite types of content as, on the one hand, teenage movies and "basic rom-coms," by which she seeks to "switch [her] brain off," and on the other, documentaries through which she "want[s] to learn something." Accordingly, she divides her information-seeking efforts according to her mood and desired viewing outcome: "It really depends on what kind of film I want to watch. If I want to watch a chick flick, I'm not going to bother about critics because they're not going to like it and I'm just there to forget about what I was thinking about and just enjoy the film, really, because it's a light-hearted film. Whereas if I'm going to watch a nature documentary or a more story-informed film or series, I might look at a review." For the latter, she reported often seeking out reviews after watching the content in order to "see if there's some sort of truth behind it, or if there are like little things that they've missed or are scientifically wrong whereas, for the David Attenborough [program] in specific I didn't, because I felt like I had that knowledge myself."

The comments suggest that Participants 20 and 22 place lower value on the merits of individual films and series while viewing Netflix: less is at stake because of the much-reduced opportunity costs. Therefore, if the content turns out to be less than satisfying, they can simply stop streaming and return to the menu to find something more suitable. (Whereas with cinema viewing, they have traveled outside and paid for a ticket to see a particular film at a particular time, rather than a monthly all-you-can-watch home-viewing subscription.) This is the contrast implicit in the comments of Participant 20, and across the interviews: although some cited examples of group VOD viewing or bingeing, almost all associated VOD with more solitary, non-social, time-filling behavior, unlike the theatrical experience, which was more often targeted for a specific film or, above all, a particular social event. Unlike cinema, VOD is there to "switch [her] brain off" and "forget about what [she] was thinking about and just enjoy the film."

In general, however, the overarching theme of recommendation use across the study revealed (1) a fundamental incomparability of source types and (2) repertoires of source use channeled via an individual, multistage, iterative process. The participants' comments in the interviews fundamentally nuanced and complexified the quantitative results in this way. For example, the survey suggested that advertisements represent one of the most prevalent means of influencing choice: 24 percent of UK residents and 29 percent of US residents said that ads influenced their selection of films or series, meaning for example, it was the third-highest used source among the US sample. However, the interviews revealed that—although many participants admitted that billboards, posters, television commercials, trailers, and online ads inflected their choices—such media exerted a much different influence than other sources such as critics' reviews. Several of the participants reported that advertisements were decisive if they already possessed considerable knowledge and a strong opinion about the content. For instance, Participant 34 gestured to the example of the most recent *Star Wars* film she had seen, *The Last Jedi* (2017): as a fan of the franchise, the mere knowledge (acquired via an ad) that a new sequel would be coming out was sufficient to decide to watch it. In addition, especially among light media users, there was some evidence of ads exerting a primary and decisive influence. For example, Participant 9, a woman in her late forties who took part in the study with her fifteen-year-old son (Participant 10), described how she came to watch the last film she had seen, *Johnny English Strikes Again* (2018). First, she saw the ad on a bus. Subsequently, she viewed additional ads on television and then she googled and watched the trailer online. Because she knew the franchise and had seen and enjoyed other Rowan Atkinson films, she did not feel the need to do further research (she usually goes on parents' user forums and blogs to check whether there is any material unsuitable for her son) and suggested to her son that they watch the film together.

This approach, which has its own consecutive steps (rather than any single source proving to be decisive), reveals a case whereby a one-source category (advertising) exerted a primary role in selection. Much more common among the study participants, however, was a hierarchical triage among and between source types. That is to say, in almost all of the interviewees' descriptions of how they came to select the film or series they had seen most recently, they detailed an ornate if-then process. In the first stage, some source(s) (most often ads, word of mouth, or an invitation by a friend or family member to undertake a joint viewing) played an initial role as a "trigger" (akin to what marketing experts call "awareness effects")[7] that made them aware of the existence of the film or series. Interviewees often registered this information passively or accidentally, rather than actively seeking it out. This stage sometimes contained repeated episodes: for example, see above the way that Participant 9 saw a variety of ads—on city buses and also television—before taking more decisive steps to consider watching the film.

Subsequently, in a second stage, participants often actively consulted further sources (an online trailer, critics' reviews, user comments, or information sites like IMDb or Wikipedia) to "research" or "vet" the film or series. Examining images or reading a synopsis and sometimes evaluation allowed them to determine the content's genre and often actors (and less often among the participants: filmmakers) and thus capacity for a satisfying viewing experience. Commonly, this second stage itself had a set routine order and procedure. Participants first assessed images (VOD thumbnails or online publicity), synopses, and actors. If at this point some interest still remained, they would go on to search for trailers, articles, user ratings, Rotten Tomatoes scores, and/or critics' reviews.

Many participants spoke also of a third stage, which they would undertake *during or after* watching the film or series. This involved scrutinizing critics' reviews, production information and history, accounts about the "real story" (particularly when the film or series was based on an actual event or person), YouTube vloggers' explanations of ambiguous endings, further articles, or interviews with the actors or (less often) filmmakers. Most viewers reported undertaking this third step when the film or series particularly impressed or affected them (in the positive sense but sometimes also in the sense of disappointment or confusion).

Because of the iterative nature of the process and the individual repertoires, routines, and trust weightings that participants applied to recommendation sources, decisions to watch (or not watch) content often stopped or stalled at one or more of the individual stages. For instance, Participant 15 reported that if she starts a series and finds it initially boring, she will simultaneously go online (via a second screen) to seek out user reviews and evaluate whether it is worth her time to continue watching. Several interviewees reported that when invited to watch a film together with a friend, they would first search online for information about the plot and actors and then watch a trailer. If the film "looked bad" they would calculate whether the social experience of spending time with their friend(s) would outweigh the potential of a film outside their preferred genre or with an actor they disliked. (For most participants, as we shall see, the social experience was almost always more important.)

Participants described a decision-making process that often included many countervailing recommendation indications, some attractive and others less so. This meant that the decision to watch a certain film or series may stretch out over days, weeks, months, or years. It might also involve false starts, list making, and procrastination. (Participant 33 described putting off watching a film on his watchlist for weeks: "I'm like maybe three hours of a psychological horror is not what I want to watch right now.") The rationale of Participant 34 for why she came to watch the last film she had seen on Netflix, *The Perfection* (2018), illustrates this layering of sources and decision-making. "Firstly [the film] was recommended to me [by Netflix]. I saw the trailer; it looked horrible, so I decided not to watch it. But then several days after that, I think I found it from the genre [row] or something and they used a different picture. It looked a bit more interesting, and so I started it. I could have stopped watching but then my friend said, 'It's supposed to be quite interesting,' so I just kept watching. It turned out to be very good." This passage reveals a number of different sources (recommender system/prominence on Netflix interface, trailer, genre search, image/advertising, and word of mouth). Some of the indications led the participant to avoid the film or (when she finally decided to start the film) to discontinue viewing, but the tip from a friend led her to keep watching, and she ended up enjoying the content.

In conclusion, we can say that, first, there is no compelling evidence to show that algorithmic recommender systems are the most important or most widely used form of suggestion. This is true even when looking *only* at VOD use, and even much more so given the fact that VOD use remains a fraction, even if a growing one, of all film and series consumption. The survey results indicate that real users consult a wide range of sources beyond recommendation engines; only a small proportion of UK (10%) and US (15%) residents say they rate recommender systems as one of their top-three sources for suggestions—compared with, for example, 62 percent of the UK sample and 56 percent of the US sample who said this about tips from friends, family, or colleagues.

Second, the interviews suggest that the choice of audiovisual entertainment is a much more complex process than simply consulting a favorite source and then making a single utility-maximizing choice: users talk about selection as a multistage, iterative process. Recommendations often occur over a wide timespan. Advertisements or word-of-mouth endorsements may be recalled months later to inform a selection decision. Such instances often constitute conflicting indicators: one ad may be a turnoff and the next one (or a subsequent, positive word-of-mouth occurrence) may favorably dispose the viewer to test the content and to persist in viewing even if the initial reaction is unengaging.

There is, third, evidence—from both the survey and the interviews—to suggest that even though source use is diversifying overall (i.e., for all channels of distribution), VOD offers the potential for an increasing one-stop-shop phenomenon. This may be particularly the case as Netflix and other sophisticated services incorporate several traditional information and evaluation sources (images, synopses, user reviews, trailers) into the platform itself. To be sure, this is not an unprecedented phenomenon: it parallels how television's teletext and then the onscreen electronic program guide (EPG) remediated and displaced hardcopy references like *TV Guide* and *Leonard Maltin's Movie Guide* in the early 2000s.

Fourth and finally, this conclusion comes with the strong caveat that there is no "typical" VOD use. The interviews not only reveal that individuals deploy their own unique routines and repertoires but also that each user will partake of a different routine or repertoire based on the specific viewing instance and goal, which may be looking for some quick-fix background noise when cooking, contemplating a mindless evening's entertainment, or seeking out an informative program. Unlike laundry detergent, films and series are complex audiovisual products, providing diverse benefits to individual users in individual situations.

II. TO WHAT EXTENT DO USERS UNDERSTAND THE FUNCTIONS AND TECHNOLOGY OF ALGORITHMIC VOD RECOMMENDER SYSTEMS?

There is contradictory evidence regarding everyday consumers' understanding of algorithms and their social uses and functions. On the one hand, much of the theoretical literature on the subject assumes that people are largely ignorant of the widespread use of algorithms in cultural selection and a whole host of other everyday applications, from life-or-death medical decisions to online dating services. These commentators use this presumed ignorance as a basis for theses about AI's encroachment upon, or even imminent takeover of, humanistic functions, categories, and concepts.[8]

On the other hand, large-scale representative surveys suggest more conflicting and altogether mixed results. A Bertelsmann Foundation EU-wide study released in 2019 indicated that some people have no idea about these matters: UK residents were especially ignorant, with one-quarter claiming to not even have ever heard the word *algorithm* before. A substantial plurality of about half, however, demonstrated a basic grasp of the concept and its applications.[9] A 2018 Pew Research Center report charted Americans' attitudes to four specific applications of algorithms (assessing criminal risk among parolees; automated screening of job applications; automated video analysis of job interviews; automated credit risk scoring). It found a public divided and quite skeptical about the fairness and effectiveness of computer decision-making, with 58 percent of survey respondents saying that these applications reflect human biases, although younger people are more likely to say computer programs can be designed to be bias-free (50% of eighteen-to-twenty-nine-year-olds).[10] This result resounds with my findings, as does the Pew question that determines a large majority (79%) to prefer human to automated decisions.[11] In my own survey, a relatively large percentage of participants answered "don't know" (UK: 19%; US: 24%) about whether they would be more willing to trust the recommendations of human critics or computer algorithms if forced to choose. This could also potentially signal misunderstanding or at least lack of clarity about algorithms and their functions in recommender systems.

We need to parse and nuance these hypotheses, however. Although headline quantitative findings suggest a substantial ignorance about algorithms, these studies often seek to test whether participants realize that algorithms are being used in high-stakes (but low-publicity) situations: to determine prison sentences, process mortgage applications, or diagnose human illnesses. For example, the Bertelsmann study revealed that while most Germans did not recognize more specialized applications such as disease diagnosis (28%) and job application screening (35%), the awareness of algorithms' roles in tailored advertising (55%) and news feeds (49%) was substantially higher. Overall, the study participants' unease regarding the application of algorithms over human judgment was in direct relationship to the value and importance they placed on such decisions: whereas majorities or substantial pluralities thought that computers should determine by themselves matters such as the efficient use of warehouse space (57%) or word-processing spell checks (48%), sentiment was reversed for judging the risk of criminal recidivism or diagnosing illness. Of those surveyed, 54 percent said humans alone should decide on parole and 37 percent believed that computers could make suggestions about offenders that humans ultimately decided on, whereas only 2 percent thought computers alone should decide. In turn, 40 percent preferred only humans to make diagnoses, whereby 53 percent thought computers should make suggestions that humans acted upon and only 3 percent believed computers alone should decide. Furthermore, the study found that for many algorithmic decision-making functions, people have become used to, and thus comfortable with, computational intervention.[12]

In fact, the consumer-facing applications of algorithms in recommender systems (Google search, digital ad personalization, Spotify, and indeed Netflix) remain altogether more prominent. Promotional stunts such as the Netflix Prize and the many fawning articles and reports in the mainstream media, not to mention the Netflix interface itself (Match Score) and direct publicity and advertising by the company mean that the basic functioning of the recommender system seems to be widely known. After all, my survey found that large proportions of UK (51%) and US (63%) residents reported having used Netflix. We could also examine symptomatic indications of awareness. For example, the humor of the *New York Times* satirical cartoon about Netflix's recommender system, presented in chapter 3, not to mention the many other parodies, memes, and pop-culture commentaries on this topic, depend on this basic knowledge being widespread.[13] Such evidence represents one way to address this research question's methodological dilemma: directly asking on a survey whether one understands algorithms and their role in recommender systems would yield essentially useless answers, with a full spectrum of known unknowns, unknown knowns, and unknown unknowns unable to be disaggregated and specified.

Therefore, more indirect means to assess users' understandings of algorithmic functioning and the recommender system were key to the interviews and demonstrations with participants. Overall—and this must be the overriding answer to this research question—interviewees revealed a solid *functional* understanding of algorithms as they pertain to Netflix, Amazon, and other consumer-facing VOD recommender systems. The remainder of this section pursues users' idiosyncratic ways of speaking about them, and the pathways of knowledge that these imply.

To be sure, some of our interviewees betrayed misunderstandings of recommender systems as they pertain to Netflix. One interviewee (#7), for example, seemed to imply that she believed that only the Recommended for You category was narrowcast to her, whereas other categories (Popular and Trending Now) might be completely neutrally broadcast to all users, uninfluenced by personalization algorithms. Participant 15 seemed to believe that algorithms were only features on the (phone and television) app version of Netflix (rather than the laptop or desktop interfaces). Other users (#4; #5) had apparently never noticed the "match score" percentage feature on Netflix—although this perhaps suggests a lack of deep interest in the platform's consumer features, rather than a fundamental ignorance of algorithmic functioning itself. When prompted during the demonstration about the match score, Participant 12 replied, "Yes, I see the matching, but I don't know, they are analyzing my search?" Similarly, Participant 24, after confessing he did not understand algorithms and how they work on Netflix, asked the interviewer whether it was "like Google" and that search engine's predictive functions. One interviewee in her early fifties who did not subscribe to a VOD service and said she did not use Netflix because she doesn't "really know enough about how to obtain it," #26, made similar comments:

PARTICIPANT 26: No, I'm not really aware of that happening, but does Amazon use that when they say things like, "other customers that bought what you bought, bought this"?

INTERVIEWER: Yes, that's also an algorithm, yes.

PARTICIPANT 26: That's an algorithm?

INTERVIEWER: Yes.

PARTICIPANT 26: Mmm. Does this all go back to Big Data?

INTERVIEWER: It does, yes, absolutely.

PARTICIPANT 26: [*Laughter*] There's so much information being gathered all the time, isn't there?

In general, such responses suggest a spectrum of dormant or *non-active* understandings of algorithms: these users are not actively pondering or considering the algorithmic ramifications of film and series suggestions during their use, but remain generally, albeit passively, aware or readily able to intuit algorithmic functioning when prompted.

Indeed, the interviews disclosed an awareness of recommender systems that remained sometimes selective and hardly absolute. Participant 7, for example, spoke cogently about the recommender system in a way that clearly showed she understood its basic functions. In other parts of the interview, however, she used language that implied that she selectively ignored, forgot, or at least downplayed the agency of its role in presenting content: "Netflix offers this program, so I jumped into it." Using the word "offers"—rather than recommends, suggests, pushes, or promotes—represents a passive articulation of Netflix's role: media products simply there, like fruit waiting to be picked off of a tree. Participant 19, who, as we shall see, was one of the participants who said she actively attempts to avoid or outfox personalization algorithms, mentioned she tried to bypass Netflix's recommender system by using YouTube, before the interviewer reminded her that YouTube also deploys algorithmic recommendation.

Nevertheless, such misunderstandings must be put into a larger context in two ways. First, the interviews revealed a certain mild ignorance of *all* information sources. That is, some interviewees seemed to conflate journalists' articles (i.e., stories about the production) and critics' evaluations. Others misunderstood the basis of the Rotten Tomatoes score or were unable to clearly or accurately differentiate between critics' reviews, user comments, and online synopses. Participant 18 said that critics' reviews were important to her decisions but in the course of the interview it became clear she meant lay user blogs and forums. Participant 19 did not seem to know whether the IMDb score, which she said she consulted often, was based on experts or lay users; Participant 25 found out he was making the same mistake in the course of the interview. In turn, Participant 21 said he sometimes checks out "ratings" online and claimed to value critics' thoughts over lay users but was unable to say whether these ratings (e.g., Rotten Tomatoes) referred to critics or amateurs and did not seem to care to learn the difference. Scholars, in their efforts to diagnose Netflix's secrecy surrounding its recommender system's proprietary algorithms and what ignorance and other pernicious effects this may cause, should be careful not to infer that users' knowledge about the workings of legacy media and recommendation sources is any more accurate. The interviews revealed confusion or inaccurate understandings of a whole range of media mechanisms and inner workings: for instance, regarding the functioning of the traditional newsroom and differences in article formats like capsule reviews, features, think pieces, and interviews; editorial separation between critics and reporters; marketing ploys and press junkets; whether trailers are commissioned and designed by producers, distributors, or others. Yes, users' understanding of recommender systems is neither universal nor complete. But their knowledge of advertising, journalism, word of mouth, and other basic sources and agents of recommendation culture is hardly any better informed or more savvy.

Second, it is important to emphasize that the above misunderstandings of algorithms and VOD recommender systems were interesting but overall anomalous features of the interviews. In fact, our sample of interviewees revealed an adequate comprehension of algorithmic VOD recommender systems' basic functions: all seemed to understand that the systems predict suitable choices based on prior user behaviors (such as viewing history) and that these recommendations are presented in the form of interface layout. Indeed, both participants with strong credibility in recommender systems and those exhibiting little trust in them evinced advanced knowledge of their consumerist algorithmic features. For example, #2, who among all of the participants had the strongest belief in the system, was well aware that titles on the left-hand side of Netflix's categories were recommended more highly than ones for which she had to scroll right.

Others revealed their knowledge with jokes or other ironic commentary. Participant 17, who on Netflix usually looks for an attractive image on Recently Added or Original Content before sampling content, said "I might not finish the series, but if the algorithm has got me beyond the first episode, sure." Participant 1, when demonstrating his Netflix selection process and revealing that he did not subscribe to the service (he piggybacked on the account of his brother's former housemate who "doesn't use it anymore"), said "I think most of the algorithm is now aligned to me." This was a common form of speech among the interviewees. Whether "the algorithm is telling me that . . ." (e.g., #21) or "the algorithm is just very confused" (#20), several deployed a mode of discourse that anthropomorphizes the recommender system, its mathematical functions, and its interface.

By way of conclusion to this section, the question of the extent to which VOD users understand the mechanics of algorithmic recommender systems such as Netflix's deserves a somewhat mixed answer. To be sure, neither the surveys nor interviews indicate that many laypeople have a solid knowledge of the informatic particulars of algorithms and the computational skills needed to design them. Very few will be familiar with the intricate terminology, nor can a large majority even define what an algorithm is (a common and somewhat trick question contained in many of the surveys cited above). And yet a basic consumer-level comprehension of algorithms' roles in VOD abounds, whether this information derives from advertising, news articles, or conversations with friends and family. Much of this knowledge surely comes from the VOD providers themselves, which, as demonstrated in this book, put forward these aspects of their recommendation engines as USPs in publicity and press releases. Filtered via tech-press and news-media confederates, these ideas trickle into laypeople's folk understandings. Indeed, among the interviewees, all had some awareness of algorithms and recommender systems, a kind of knowledge that seems to be built into the consumer experience of especially Netflix and its most prominent competitors such as Amazon Prime. (Even those few interviewees who reported more or less never watching VOD seemed to recognize algorithms or at least individualized recommendations as a feature.) Best described as a selective, non-active, and functional knowledge, real users' folk understandings sometimes reveal a sort of cognitive dissonance: participants, when speaking in the abstract, knew that their recommendations emerged from a personalization algorithm at the same time that they deployed language that suggested these programs were somehow just "offered by Netflix." This is a consumerist folk understanding, not scientific epistemology, and thus it often mixes in myths and false assumptions.

As a final note on this section and a way to segue to the next, it should be said that some interview participants exhibited a way of speaking that readily understood basic algorithmic functioning and yet remained uninterested in it, what one might call the "don't-care" faction.[14] Participant 11, for example, confessed "Yes, I think algorithms have a big impact on what I want to watch." He acknowledged their potential effect without registering concern or even expounding or reflecting further, besides comparing "good recommendations" (in this case YouTube), which often lead him to watch clips for hours, to less effective ones (Netflix, Amazon). Perhaps the most exemplary respondent in this vein was Participant 21, an infrequent VOD user (about once per week) from Scotland. When asked his general opinion on algorithms, their accuracy of recommendations, and their overall effects on his film and series viewing, he replied: "I'm aware of them. I don't think of them. I don't know. I don't care.... Yes, I think it's taking away people's freedom of decisions in a way. It's feeding them what it thinks they want to see and they're not going to explore a wider variety of things. I wonder how much of what I watch has been fed to me by an algorithm rather than me consciously deciding." When asked in a follow-up whether he found that state of affairs frustrating, he replied "I don't care. I don't think I watch a lot of it for it to have made a really sophisticated algorithm for me [*laughter*]." This participant, who in general has a low interest in films and series beyond what he likes to discuss with friends, also maintained little interest in the functioning—algorithmic or otherwise—of Netflix and its recommender system. "I've never actually paid attention to the match thing. I guess that's the algorithm telling me what . . ." [Interviewer: "Exactly."] "Right, okay. I've never paid attention to that. Knowing that now I might pay attention to that. I don't tend to read the descriptions much. A lot of the time I put something on for five seconds, get bored, put on something else for five seconds. Because there are so many of them it's hard to—[I've got] less attention span to be able to stick to one thing when there are so many options." The mere fact of knowing did not necessarily lead Participant 21 to reflect on this use or make substantial behavioral changes. This suggests that—for those scholars deeply concerned about this topic—gauging knowledge, as measured in "awareness" or "understanding" of algorithms and recommender systems, may not even be the most appropriate or dispositive questions to pose. Fully aware of the potential downsides of using algorithm-led applications, many people simply do not care about them or their potential effects.

From the perspective of someone who has devoted several years of his life to studying this subject, it remains a simple task to point out instances where real users misunderstand or misinterpret publicly available information about how algorithms, Netflix, and other VOD recommender systems work. Such an exercise, however, would miss the important point here. To paraphrase Toff and Nielsen, VOD users' folk understandings and theories of recommender systems may not be factually accurate, but they are social facts.[15] Real users' ideas about what is or is not personalized or determined by algorithms or controlled by a VOD platform's parent company, (mis)understandings infiltrated and informed by publicity and journalistic discourses and peppered with urban legends and conjecture, may not be technically accurate.

But they remain at least true and real as lived experience. These assumptions and understandings guide their users' actions on the platforms and, in turn, inform their perceptions of their own taste and identity as film and series consumers. As we shall see, some see VOD platforms and recommender systems through the prism of filter-bubble discourses or cultural anxieties about data surveillance. Some even speak in a manner aligned with Netflix's perspective and thus partake of an industry discourse: according to #22, Netflix's trailers "are a good marketing tool, so that's cool for them." Overall, however, most participants understood algorithms and recommender systems via a consumerist angle: whether or not the algorithms were working for *them*. This is a *functional* folk understanding, a mode of seeing and speaking (informed by but) unlike the many other discourses on this topic: those common among computer scientists, psychologists, business observers, media scholars, and so on.

III. CREDIBILITY AND EFFECTIVENESS

The quantitative surveys—both my own and those conducted by the Pew Trust and Bertelsmann Foundation—might lead one to believe that there could be a small but stubborn minority who maintain a blind faith in the recommender system, rating its credibility and effectiveness as high. After all, according to my survey, 7 percent (UK) to 12 percent (US) said they trusted computer algorithms more than human critics; depending on how the question was asked, 7 to 13 percent of UK adult VOD users and 9 to 14 percent of US adult VOD users said they were likely to use the recommender system to select content. A few select demographics (US Hispanics: 17%; UK eighteen-to-twenty-four-year-olds: 17%; US eighteen-to-thirty-four-year-olds: 21%) were relatively more keen to say they trusted algorithms over critics. These small but significant minorities of the population seem to maintain credibility in the effectiveness of the systems. Furthermore, as examined in this book, the theoretical secondary literature implies or outright claims that there is a silent majority who simply follow algorithmic suggestions.

The interviewees, among whom there were several who said that they were likely to use a VOD recommender system to inform their content choice, provided a much more nuanced picture. None of the participants seemed to place blind faith in recommender systems. Indeed, only four participants (#2, #8, #10, and #31) attested to strong credibility in the recommender system at all. Participant 10, a fifteen-year-old, spoke of recommender systems confidently and much in the vein of industry discourse: "It encourages new shows, it also just encourages you to choose wisely what you're going to watch, because there's way too much choice." The example of Participant 8 is also revealing (and very exceptional) in this regard. She was one of the very few participants who not only read critics' reviews but could also name a critic she followed; she detailed frequent word-of-mouth instances with her mother and her colleagues. Nevertheless, later in the interview she entertained the thought that algorithmic suggestions might be more accurate than friends' word of mouth: "Because the [BBC] iPlayer and [the VOD service of] Channel 4 learns what you like, doesn't it? Over time. Whereas your friends don't always necessarily know what you like. I think it's quite different then, because the apps have learned what I like, but my friends just pick what they want to see and throw it out there to everyone and see what happens." (We must unpack this statement carefully. Although it could imply that she trusts recommender systems over friends' tips, it could also indicate unsatisfying experiences of having friends choose content to watch together without properly taking into account her tastes. Conflating word of mouth with joint decisions or invitations was another iteration of benign ignorance exhibited among some interviewees.)

Participant 2 expressed a strong belief in Netflix's capacity to pair her with suitable content. When asked whether using the streaming service has increased, decreased, or had no effect on the diversity of content she consumes, she replied that she watches "more diverse type things because of Netflix. Yes, exactly, because there are certain genres and movies that I don't really like, but you look at—it comes in your face and it's, like, 'Oh, let me just click and see,' and then you're stuck with it, so yes." This response suggests that the participant's high degree of trust in the recommender system trumps what would have been a negative judgment of the content based on her impression of the title and image. The following exchange—from her demonstration of how she chooses content on Netflix—is revealing in this regard:

INTERVIEWER: So, let's scroll for a bit. Let's take the time, and let's just search for a particular movie you would pick right now.

PARTICIPANT 2: Yes. I would pick the *Queen of Katwe* [2016].

INTERVIEWER: Right. So, this is in a category "Movies Based on Real Life." It's the first option in the category. You often tend to go for the first option. That has happened a couple of times already.

PARTICIPANT 2: Exactly, I do. The first option is—more of them match to you, you know. So, if you look at the match [score], it's 96 percent of the things I like.

INTERVIEWER: Oh, right, so you look at the match, too?

PARTICIPANT 2: Exactly, yes. So, it's something that is, like, a trend. So, it will give you that.

INTERVIEWER: So, the algorithm is having a huge impact on your decision process?

PARTICIPANT 2: Exactly, yes. So, it will just make me want to click, and look at it. If I look at it . . . .

INTERVIEWER: So, what if the *Queen of Katwe* had a match of, like, 65 percent?

PARTICIPANT 2: 65 percent?

INTERVIEWER: Yes. Would that have worried you?

PARTICIPANT 2: Well, yes. I mean, exactly. I would still look at the storyline to see the excerpt. If it's something that I would want to know, then I do, but if I look at it and it's not something that appeals to me, I just skip.

Even in this excerpt—and with this subject, who of the thirty-four participants most loyally spoke of the Netflix recommender system—belief in its power is not entirely absolute. It is clear from the exchange that despite saying the match factor is most important for her, in fact she would have perhaps given the low-match content a chance because of her potential overall interest in the image, synopsis, and trailer. Among our sample of interviewees, there was no apparent correlation between a technical or consumerist understanding of algorithms, on the one hand, and trust in or use of the system, on the other. Those with the most sophisticated knowledge of the system were split on their trust in it, and the same was also true among those who seemed to know much less about the mechanics and functioning of VOD recommender systems.

Participants 2, 10, and 31 were, among all of the interviewees, the most forthright in their admiration of tools such as Cinematch. On the other end of the spectrum, some participants rejected VOD recommender systems outright: "I'm old-school" is how Participant 9, a woman in her late forties, rationalized her nonuse. A few other of the older interviewees who did not subscribe to, and rarely if ever used, VOD (e.g., #26, #27, #28, and #30) made similar statements. Perhaps most succinct and vehement are the words of Participant 33, who expressed his belief that VOD recommender systems uniformly provide poor recommendations and that algorithms do not—and cannot—understand his taste. He said, if forced to choose, he would "absolutely" choose a suggestion by a human critic over one provided by an algorithm. When asked why he does not use VOD recommender systems to guide his choices, he replied: "because it's an algorithm and it just suggests films based on an algorithm that doesn't work. Because the films they are suggesting are absolutely not what I would watch." Participant 33 revealed himself to be almost offended by the suggestion that someone (especially him) would want to follow an algorithm when choosing content. This reaction—quick and informed, indicative of someone who had clearly considered such issues at least informally before—occurred among very frequent viewers with high confidence in their own film and series tastes (e.g., #32). Nevertheless, this correlation was not absolute: Participant 31, despite heavy media use and high confidence in her own taste, adopted a diametrically opposed attitude to recommender systems' suggestions.

The discussion of this question thus far shows how at least some users have strong, well-developed beliefs in the potential of recommender systems, both on the extreme positive and negative ends of the spectrum. Nevertheless, among our sample, a more ambivalent view was much more common. And even those participants with a high degree of faith in the recommender systems suggested that their belief was not absolute. Other indications, such as the thumbnail image, tempered their trust, and many registered suggestions "with a grain of salt." The national survey results indicate that only a relatively small proportion reported being likely to use the recommender system to select films and that an even smaller minority say that the recommender system figures among their top three sources most likely to influence their choices. How do we reconcile these preliminary indications?

As a way to explore this phenomenon, the interviews asked participants in what situations, how, and why they used a VOD recommender system—or, if they reported not following such recommendations, why not. Despite a few absolute pro or contra answers, illustrated above, ambivalence prevailed among our interviewees. Three prominent and overlapping answers dominated, attitudes or folk theories that I subsequently explore in turn:

(a) Weighing poor recommendation outcomes or security concerns against convenience and other consumerist benefits, summed up in feelings such as, "It's a useful/convenient tool, but . . ."; or "There is something cynical in me that mistrusts them . . . but as a consumer it's very important";

(b) Beliefs about recommender systems misjudging moods and being inadequate to understand the intricacies of human taste (e.g., "It doesn't know how I'm feeling that evening");

(c) Feelings about surveillance, filter bubbles, and a lack of variety in recommendations, divided by a belief in fate ("there is no way you can run away from [algorithms]") or free will ("you can just go to the different genre").

A. Weighing Theoretical Mistrust against Consumerist Convenience

Confirming prior studies, in particular Novak's small-scale examination of millennial university students, the interviews revealed respondents to be generally dissatisfied with Netflix's and other VOD platforms' recommendations.[16] The complaints about "bad recommendations" and the reasons for their low trust in or infrequent use of recommender systems were varied. Some interviewees, as we shall see, bemoaned frequent suggestions of the "same" content. Overall, attitudes hinged on a simple mismatch in taste. Respondents typically concluded that although recommender systems are sometimes accurate, just as (or more) often suggestions proved to be far off the mark. When asked about Netflix and Amazon recommendations and why she did not use them often, Participant 34 said: "Half of the time I've just got zero interest in any of those films, so it's just pretty off." For Participant 16, "Sometimes, they are not correct. Definitely, because I don't know why it's even there." This was also the general opinion of Participant 20:

> I think [algorithmic recommendations] can be useful, I'm not necessarily scared of them. I think, sometimes when you just don't really want to think about what you're going to watch, it's nice to look at a suggestion, but I always tend to check the trailer before I start watching it. And then, yes, sometimes they're spot on and I really, really like the film, and sometimes they're just a bit off and it's not really what I like to watch, but I understand where they're coming from, but it's just a bit . . . yes, just outside of what I like.

This last statement reveals a theme throughout the interviews: respondents reported that their acceptance of suggestions was, first, highly situational, and second, verified via other information sources: trailers, user ratings, reviews, and so on.

For several participants who responded in this vein, using recommender systems over a period of time has resulted in a gradual loss of credibility in their capabilities. Participant 13 described his initial belief in the recommender system to provide appropriate content and then a subsequent erosion of trust in the system's effectiveness because of repeated poor tips:

> I would say, when they say "because you watched this, we suggest this," I see it and I go sometimes. In the beginning, I believed what they said, so [I thought], okay, this is something. Because I liked this movie, okay this will be—this suggestion will be good. But I figured there one day, their suggestion is not really close to what I watch, it's kind of like different stories, like not really even close. Sometimes I can't put them in the same category. I would say, before, yes, when they said, "because you watched this, you should watch these movies," I believed that. Then now, no, their suggestion, I never pay attention like before.

In his statement, this participant suggests he might have more trust in the system if he knew more about *why* the films were being recommended, an issue (as we have seen) that computer science and psychology literature regarding recommender systems and persuasion overall highlights as an important user-facing consideration.[17] But overall, his comments and the others above suggest that whatever credibility algorithmic systems may have had initially and in the abstract to users—evoked no doubt by marketing promises and fawning reportage—many have subsequently determined such powers to be lacking in the practice of use. As such, they rate the systems to be at best "half" effective, an overall ambivalent response.

A second subgroup diverted questions of use and trust in recommender systems to complaints about Netflix's and other VOD systems' poor usability and selection, and above all, the recommender systems' failure to satisfyingly manage choice. (It is important to note that this response is somewhat different to the response above that the recommendations were "inaccurate" or not trustworthy.) As discussed in this book, VOD recommender systems accrue functional meaning only in the context of their underlying catalog. Because consumers who have too many choices risk feeling overwhelmed (and not choosing anything at all), recommender systems seek to predict and present a manageable subset of suitable choices based on prior viewing history and other factors. Several of the respondents implied that Netflix and other VOD platforms had essentially failed in this objective. Participant 3 complained about too much information or stimulation on the interface (many if not most of the interviewees volunteered that they disliked Netflix's automatic trailer function) and Participant 5 reported the amount of content on Netflix to be "overwhelming," which resulted in excessive searching time, despite the recommender system's supposed simplification of content presentation. In turn, Participant 8 said she let her Netflix subscription lapse because executives "oversaturated" the platform's catalog "with their own content [Netflix Originals] and I spent more time looking for something to watch than actually watching anything, so I just got rid of it." She said she prefers curation and small menus, also in restaurants, and favored providers like BBC iPlayer and All4 because, compared with Netflix, they proved easier to navigate and find suitable content. Participant 9 was more ambivalent about this matter, noting conflicting feelings of overstimulation and free choice: "It really does take time, sometimes, and I spend time taking note of a few and not so many. . . . I still like it and I like it because you have a lot of options. Sometimes, when they know what you like they advertise it to you, and they recommend it to you in emails." Beyond these examples, which arose in interviews specifically in conjunction with the recommender system, there were a whole host of spontaneously volunteered invectives about Netflix's, and even more so Amazon's, poor selection; these ways of speaking should be seen as of a piece with one another.

A third subgroup expressed their ambivalent view on consumerist benefits not only when grappling with feelings about a lack of accuracy or manageable choice, but also with another potential downside: security. A relatively light VOD user (Netflix once per week; Amazon once per month), Participant 22 epitomized this view. When asked whether she felt the suggestions were useful and accurate, she replied:

They definitely help you to make more selections, I think. Yes, I think they work well. I think they could be a little bit more accurate, but I don't know how though. Still I think they have been working very well for me, and I've been relying on using it very frequently. Yes, algorithms for me are a good thing. I know probably security-wise and all of that it shouldn't be like that, but as a consumer it's very important for me now on Netflix that they will actually give me less time to select something. Because sometimes it takes you ages to select what you want to watch. And they [help] like that and will tell you the list of the movies that you might be interested in. You go directly on that, and then based on that you can select what you want to watch. Yes, I like algorithms.

This statement illustrates a respondent weighing moral and security issues against utilitarian concerns such as entertainment, enjoyment, and ease of selection. Saying "I know ... it shouldn't be like that" acknowledges the social pressure she senses (through the news media and perhaps through the social interaction of speaking to a scholar in interview) to take privacy and surveillance concerns seriously. In her case she ultimately concludes that, on balance, the consumer functions are more important.[18]

Participant 17, in contrast a very heavy user (daily Netflix use, also heavy use of cinema, DVDs, and linear television), nonetheless reckoned with a similar calculus: balancing a general awareness (and mistrust) of algorithmic functionality, attendant data protection concerns, and feelings of overwhelming choice (à la #3, #5, #8, and #9 above), against the ease of functionality and the convenience of exposure to new titles.

> There is something cynical in me that sort of mistrusts [algorithms and recommender systems]. But, at the same time, I think it's nice to see what's out there, you know, and what people are watching, so that maybe I can watch it as well. If it is popular and recommended, I might give it a go. But also there's just so much to go through. I think unless you are looking for a specific program, or like a thumbnail catches your eye, you are going to have a pretty slim chance of finding something. To me anyway. There is almost an excess of things to watch, and sometimes it's just quite overwhelming. I've got so many things on my watchlist, and I think, "I'll get to that one day. Everyone goes on about that series." It's just been on there for two years, and I just haven't, whereas new things I tend to watch quicker, I suppose.

Her final judgement was to avoid the recommendations on balance. Indeed, as a maximizer who consumes lots of information before deciding what to see, she uses the recommender system much like other participants used ads or word of mouth: as an initial first-stage "trigger" that made her aware of a certain title, rather than as a serious suggestion she would follow-up on. "On Netflix? Well, it's difficult because I suppose I tend to avoid [their recommendations], because I tend to check things out when I have been recommended them [by word of mouth], or if I have seen something on Twitter. Very rarely have I actually clicked on something that I've been flicking through for like an hour, you know? I'm sort of trawling through the mire of things that are available, and it's almost too much. I suppose the algorithm works against me because it puts me off if anything. There is just so much." Note here the way that her knowledge that an "algorithm" was making the suggestion left her less likely to follow it. We will see other examples where users make such explicit trust distinctions between human and computer recommendation sources.

In sum, these participants consider a recommender system primarily for its consumerist applications and implications. In particular, they deem the tool to at least partly fail to achieve some fundamental objectives: to furnish suggestions that match the taste of the user, to provide a meaningful and suitable quantity of choice on an interface that is pleasant to use, and to do so in a way that inspires confidence in security and engenders overall trust. The comments in this section show a wide spectrum to the way that individual users ultimately wrestled with these downsides in light of the perceived convenience of the system: for Participant 22, security issues were at best a theoretical concern, whereas Participant 13 seemed altogether more disillusioned and suspicious. Although this group of interviewees sometimes implicated their emotions in their speech (feelings of being overwhelmed or overstimulated or wronged by poor recommendations), this topic often arose in a somewhat detached way—that is, as a technical and business problem that VOD providers could potentially solve with better design and better content-acquisition policies. In contrast, the following folk theory squarely implicates individual and supposedly unique human characteristics: tastes and moods.

B. Algorithms Misjudge Moods, Taste, and Feelings

Promotional statements suggest that Netflix is developing the capability to not only ascertain general genre preferences, but also anticipate viewing moods. According to CEO Reed Hastings, "We hope to get so good at suggestions that we're able to show you exactly the right film or TV show for your mood when you turn on Netflix."[19] A second attitude, prevalent among the interview sample, implied that Netflix and other platforms were failing miserably to achieve this aim. Recommender system suggestions may or may not be accurate, these respondents believed, but because the tools misjudge particular moods in particular viewing situations, they have an at best middling use value overall. This sentiment, expressed along a spectrum of mostly negative but sometimes cautiously optimistic opinions, involved the folk theory that either algorithms can *never* understand human tastes or algorithms cannot fully understand human taste *now*, but will do so in future.

Participant 3, when asked during his Netflix selection demonstration what he would choose and why, revealed that the position of the content on his screen and the match score may have played a role in his selecting *Élite* (2018–present) on this occasion, but it was hardly dispositive: the trailer and images were much more important in his decision-making. His response readily acknowledged that an algorithm was suggesting the content based on his prior viewing behavior. "For me, I would watch the trailer and read the description. I think the last thing would be the match [score], because I wouldn't choose a film because it matched my taste.... General matching can't capture my current feeling in that moment." When asked if he trusted the algorithm, he replied that "I think in general I trust the recommendation, but I think it would be my last criterion. The order would be to see the trailer and then read the description, then look at the match recommendation." Part of his ranking the recommender system's match score as relatively unimportant derives from an ingrained belief that algorithms cannot decipher (his) moods.

The comments of Participant 5 are similarly illuminating: "Sometimes I get good recommendations, but it's interesting; sometimes, if I'm in the mood for a chick flick, I watch one and then it recommends lots of chick flicks for a while, and then I'm not in the mood for that. It puts me off a bit. It gets stuck with one particular thing you watched and then it recommends a lot of those. Sometimes you want a greater variety." For this respondent, the misjudgment of moods undermines her overall trust in the system. According to Participant 16, who articulated her problem with Netflix in a similar way, the systemic failure to identify appropriate moods and tastes led to more work, searching, and sources: "Basically, I try to look to more options, just not the things that are recommended to me. Sometimes, I think they leave things out because Netflix doesn't know actually what other things I like. For example, I usually watch really easy-going movies like comedies, chick flicks, or something like that, so, obviously, these are the kinds of things that it recommends to me, but, sometimes, I do want to watch a good movie with a good story and something interesting. Now, for example, Netflix doesn't recommend it because I don't usually watch those kinds of films there."

Participant 32 further illustrates this attitude. When asked to provide her thoughts about VOD suggestions, she said:

> I don't think an algorithm can articulate nuances with taste. It doesn't know how I'm feeling that evening. So, I'd say it's a good ballpark . . . kind of general guideline. There's actually a website that's really good for it where you can type in any—it does it for music, so, I'm just like, I like this thing, and it will come up with a whole bunch of stuff you might want to check out. Yes, it's okay, but not perfect. I would say Spotify is better in terms of music for

recommending bands and stuff [than Netflix is] with films. Again, I think it's just a question of tone, because it's easier to convey with music.

For this respondent, her lack of belief in computers' capacity to apprehend but also comprehend human moods and tastes meant that no VOD system—whether Netflix or any other now or in the future—could ever be truly effective because of a fundamental incongruity between the world of humanistic taste and computational calculation. This represents a lay iteration of the Hallinan and Striphas "algorithmic culture" thesis: human taste and cultural transmission are incompatible with the world of numbers, machine learning, and artificial intelligence.

In contrast, Participant 23, a woman who works in a computer science field, deemed the VOD recommender system important. She displayed a strong confidence in computational and algorithmic systems in general and reported using recommender systems as an initial guide in her personal triage, before always double-checking recommended films with user scores on IMDb, which she finds "probably more accurate." Nevertheless, she expressed her disappointment with the device's practical value, in particular for the way it misjudges her viewing moods:

> I'm quite interested in the way how they code Netflix [because of my] background, so I do tend to actually check what is the score of the movie or of the series that they are offering. To be fair, I don't really find that quite accurate sometimes. I do find some of the films are a 60 percent match. I read the review and I watch them, and they will be actually a 90 percent. Sometimes I find it's a rom-com that is not something that I like, really, or I enjoy. They offer them as a 90–95 percent and they're not, so I don't know if they are based on the things I watch or they are based on, I don't know, my age or my gender.

This comment reveals a theme among participants, who often wondered whether they possessed an innate characteristic (particularly identity factors like gender, age, or ethnicity) that could explain why they received a certain recommendation: for instance, "Is it because I'm Jewish?" Such responses symptomatically reveal a certain latent credence is being given to the system. They essentially accept that there must be *some* truth about the recommendation and that therefore some aspect of the suggested content (national origin, genre, identity of protagonist, intended audience appeal) *must* have a significant relationship to some aspect of their own identity (national or regional origin, ethnicity, race, gender, sexuality, class) or some aspect of their taste that they might not be aware of themselves. In essence, these respondents have internalized the beliefs expressed in the *New York Times* and other middlebrow publications' feature articles, that algorithmic recommender systems are able to understand the deep structures of taste better than humans and can reveal to individuals more about their tastes than they know themselves. For a symptomatic pop-culture expression of this sentiment, see Lizzie O'Shea's programmatic *Guardian* column, "What Kind of a Person Does Netflix Favourites Think I Am?," in which O'Shea rejects the identity that her Netflix recommendations ("a listicle of romcoms featuring a hapless white lady, with a kind heart and an unfortunate failing, struggling to get it together with the square-jawed-but-equally-flawed male lead") superimpose onto her.[20]

Participant 23 was one of several interviewees (e.g., #32) who spontane-ously and negatively compared the Netflix recommender system to Spotify, which they deemed more responsive to moods:

> I prefer the Spotify recommendation system, which is basically they have, "How do you feel today?" For example, I have days where I do want to listen to Latin music, but there are days in which I also need to study and I'm not going to put Reggaeton on. . . . I would prefer Spotify to tell me, "Okay, is this a day where you're going to focus or is it a day where you're feeling a little bit low or is it a day when you're going to relax?" I can go in there and actually, according to how I feel, I will pick something different. With Netflix I do find it quite difficult to find new content because I do have to research it a lot because on the homescreen the only thing that appears is exactly the same shows as *Friends* [1994–2004], for example.

Her objection to the recommender system, in other words, revolves around its faulty and obstinate obsession with recently watched content:

> Netflix is just—if I watch *Friends*, they tell me and expect me to watch *Friends* for the rest of my life. It doesn't consider that I may want, one day, to watch a documentary. That's another thing I find. If I watch a documentary, let's say, today and I go to Netflix tomorrow, my home screen is just going to be full of documentaries now. You're like, "Okay, because I watched a docu-mentary yesterday that doesn't mean that I'm going to watch the same [thing] that you are offering me now." I feel like it just sticks to your last thing that you watched.

Scrolling to more specific categories was not an option, according to this par-ticipant: "I do feel, again, that that's just showing me exactly the same. I don't know why. I feel like it's showing me a 'Recommendation for [Participant 23]' and then a bunch of the same movies appear in a list. Again, I do have trouble with the Netflix recommendation just because I think it's exactly the same in every single category." Such complaints about the recommender system's inability to estimate moods often bled into larger grievances about "sameness" and a lack of variety, as we shall see a major topic across the respondents.

Some participants attributed the recommender system's poor assessments of moods to their own supposedly unique taste: in other words, they asserted that their own particular preferences remained too eclectic for machine learning to encompass or understand. Participant 20 characterized VOD recommender systems overall as useful—she uses it often when in a "browsing mood" of viewing on Netflix, watching the in-house trailer and taking a peek—but attributed poor recommendations to algorithms' inability to comprehend her one-of-a-kind taste.

> I think the algorithm sometimes pushes me to watch more of the same thing, but I also think that the algorithm is just very confused with my taste, because when I'm on Netflix I can watch a teen rom-com, *Top Gear* [2002–present], and then I watch a nature documentary, which is three very different things. So I also think the algorithm is sometimes just a bit confused with what I want, and therefore I still get quite a wide variety, because the people who . . . watch just one genre, as opposed to three, usually stick within that one, whereas I just sort of . . . I get suggestions from all three of those groups of people.

This sentiment—which understands one's own taste or moods as being eclectic, diverse, or omnivorous and implies that "others" have more unified, monotone, or easily decipherable preferences—was a common subtext among several participants (e.g., #21). This belief was used to justify considering the recommender system to be overall useful (for certain moods and viewing situations, or for "other people") even as they complained about suggestions being off the mark or inappropriate for other frames of mind. The speakers seek to link taste and moods—which they imply are fundamentally human and unique to individuals (especially themselves)—to a hermetic realm that Netflix and other VOD recommender systems cannot successfully penetrate. It is important to note that this view occurred on a spectrum: whereas Participant 32 implied that no algorithmic recommender system could ever perform this task (even at the same time offering Spotify as a contrasting, positive example), Participant 23 saw this essentially as a design flaw that developers could fix.

Even further toward the end of this spectrum was Participant 31, who evinced an overall strong credibility in the recommender system and took the most positive view on this topic: although admitting the system was fallible, she posited that she could "correct" the bugs in the system through her own agency and free will. When asked whether she believed the Netflix recommender system provided her with good recommendations, she responded: "I think it really comes back to myself and my own algorithms. Like, I trust that I've watched enough on there [for my algorithms] to be able to recommend something that I would also like. Also, the percentage of match—I rarely get, like, a middle ground. It's very often very much, like, 90-plus. . . . So I trust that they've not just, kind of, lumped things in." Even this user, with a relatively high level of confidence in the system, revealed that she did not simply follow every recommendation, but rather evaluated it against what she knew about the film or series, taking the suggestion into account, but hardly following it slavishly: "I mean, I think [Netflix algorithms are] still learning in the sense that sometimes—because I know, like, for instance, I'm not going to like *It's Always Sunny in Philadelphia* [2005–present] just because I watched *Arrested Development* [2003–2006, 2013, 2018]. Like, you know, some things, I think, they just tie in because of supposed genre, but they're not actually the same thing at all. But I do [trust it], overall, you know. I think I can take it with a pinch of salt, but I do trust what they're saying." When asked why she listed recommender systems as one of her top three sources, the interviewee shared the belief that the algorithm—although not yet perfect—remains in the process of learning her taste, still under engineers' careful development, so that one day, it *will* give her suggestions that correspond more perfectly to her preferences.

This set of beliefs is revealing for several reasons. This participant essentially shares the complaints of other respondents (and indeed almost everyone in the study): the recommender system is not (yet) able to always provide suitable suggestions. Nevertheless, her basic acceptance of the system's integrity leads her to consider these flaws as a matter for users to deploy their own discernment and discretion. Note the way she calls Netflix's personalization features "my own algorithms": for her they represent unfinished devices that the master user can (and must) train, rather than tools designed by others to which she is involuntarily subjected. Note also the way that, as a heavy and frequent film and series consumer, she intimates that her own prior knowledge of film and series culture prevents her from actually succumbing to bad recommendations. Even if suggested, she will never watch *It's Always Sunny in Philadelphia* because, based on her own foreknowledge beyond the platform, she understands that it does not suit her taste. Rather than a helpless victim, Participant 31 would see herself at fault were she to take up such a suggestion; it is *her*—not Netflix's—obligation to filter out poor recommendations. Finally, it is remarkable how Participant 31 foresees the technology improving with time and implies that it remains her responsibility to train "her" algorithms. Overall, this optimistic reckoning gestures toward an integral theme to which I now turn: users' folk theories about whether the recommender system is a tool of control with predestined effects on the user, or whether consumers themselves have final agency and responsibility, a "free will."

C. Data Surveillance and Filter Bubbles: Fate vs. Free Will

Studies suggest that some people may be very concerned about data protection and surveillance methods in algorithmic applications.[21] One book-length effort devotes itself to the ways in which some users "resist" recommender systems, avoiding them or attempting to outfox or "talk back" to algorithms.[22] Indeed, the present volume has examined scholarship that details how media users overall do not like to feel "pushed" into content choices.

A number of interviewees spontaneously deliberated on this topic, with varying degrees of disquiet. Participant 15, for example, offered a generally benign view of Netflix algorithms, contrasting the recommender system to cookies and online tracking via internet shopping, the latter of which bothered her more. This view would seem to be in line with the results of the Pew and Bertelsmann surveys. Both Americans' and Europeans' levels of concern about algorithms stood in direct relationship to personal stakes: algorithms in medical diagnostics and parole decisions elicited more anxiety than those deployed in areas that ranked lower in importance to civic life.

A few respondents, however, did maintain more serious concerns about VOD recommender systems' data surveillance capabilities. One aspect of these responses was a visceral reaction against feeling "categorized" or "forced" into a taste silo, an antiauthoritarian sentiment that expressed anxious perceptions of lost control. Participant 16, in an echo of the aforementioned "it's a convenient tool, but ..." folk theory, said she found the match score important and praised the consumerist convenience of the interface. At the same time, however, she voiced fears of missing out (FOMO) because she imagined the recommender system hiding potentially satisfying content from view. She said she thinks the recommender system is "good because, usually, the content that I would like would come first based on what I've been watching. That's good because I don't have to be looking for things." As convenient as this function is, however, she wondered in the same breath about what she might be missing: "On the other hand, I feel like there's a lot of content where I don't know that it exists because I've never seen it [on the homescreen]. It's always the same content appearing instead of changing or showing me something new. It's feeling like they want me to watch it, like they are suggesting something. It's like if it was something they were imposing instead of recommending." Participant 16 attempts to "avoid" algorithms because she feels uncomfortable about feelings of reduced agency. Her perceived (and disliked) loss of control implies that a recommendation emanating from an algorithm is qualitatively different (and inferior) than one from another source, such as word of mouth, a critic, or perhaps even advertising. In addition, in this passage she is weighing the positive "trigger" function of the recommender system against her feeling that the tool not only *offered* the same films and series—it foisted them onto her. This statement reveals an example of antiauthoritarian "resistance" rhetoric: the word "they" here ("they want me to watch it . . . something they were imposing instead of recommending") implies a corporation and a machine, a nebulous authority against which the user wishes to rebel. The participant is wrestling with a fear of missing out (on quality content not recommended to her and thus invisible on the interface), even as she ambivalently recognizes the consumerist benefits of reduced search-and-browse time.

Participant 19, in turn, attempts to "avoid" algorithms with mixed success. She broached the topic of algorithms several times during the interview, indicating some concern. When asked whether she uses social media for films and series tips, she replied, "Facebook, yes. I am guilty of this charge. Yes, I scroll quite a lot down on Facebook actually. Most of the time I'm using it to hear things about my friends, it's mostly for the things that I follow. Because of the algorithms, of course, I keep getting more about the background and storylines and other different series depending on the shows and the series. Sometimes they recommend other films to watch because of something else that caught me. Yes, I use Facebook a lot as a reference." When asked whether she follows the suggestions of the Netflix recommender system, she answered: "Yes, sometimes I do but it tends to be that I prefer to go and watch the trailer on YouTube. For some reason, I prefer to go separately because of these algorithms, as you say. Whenever you click on something it will be linked and then more offers and promotions. I really don't like that; I get annoyed by those." When the interviewer reminded the participant that YouTube also deploys algorithms to suggest content, she replied: "Yes, yes, it will happen, yes. There is no way that you can run away from them, and they're useful up to a point." Indeed, her overall stance was altogether ambivalent (if not seemingly contradictory). When asked about her general opinion on the role of algorithms and about whether she found them helpful, she responded: "They probably are. I tend to avoid them; I don't like being told what to see. I think that's with algorithms in general, with YouTube or even Facebook, I try to avoid them. I think I do follow them sometimes." Although admitting that she did sometimes accept algorithmically generated suggestions, Participant 19 seemed to chafe, in an antiauthoritarian spirit, at the idea of a computer determining her taste or her consumption choices. Like #16, agency over media selection and consumption was important to her, and she viewed algorithmic recommender systems—despite their obvious conveniences—as a potential threat to this freedom. Overall, the response demonstrates someone grappling with (and conflicted about) her free will vis-à-vis recommender systems. On the one hand, she highlighted her attempts to "avoid" and escape algorithmic suggestions; on the other, she despaired that there remain no true means to elude them, "no way that you can run away from them," and acknowledged several times in the interview the use value of the tools, if only for consumerist purposes. Resigned to algorithmic suggestions as the fate of contemporary cultural life, her folk understanding of recommender systems deems them unavoidable despite her own persistent efforts to resist. Like Participants 16 and 17, she betrays a pronounced defiance to being "told" by an algorithm what to see.

If concerns about data surveillance were a minor subtopic among the interviewees, the overall complaint of sameness and suggestions of a "filter bubble" were more serious and widespread. We must consider this trend together with the survey indications (Figure 7) that most respondents felt that the films and series they have watched since they first started using VOD had become more diverse (UK: 26%; US: 30%) in terms of genre, language, or time period, or at least stayed the same (UK: 50%; US: 47%), rather than having become less diverse (UK: 7%; US: 7%). Indeed, this phenomenon must also be considered alongside the fact that, when asked directly, also most of the interviewees—even those who would at other points in the interview complain about the "sameness" of the recommender system suggestions—said that the films and series they have watched since they started using VOD were either more diverse or just as diverse as before.

One of the participants (#13) said he did not mind filter bubbles, and another (#4) actually welcomed receiving similar recommendations among a circumscribed set of genres and styles, appreciating the efficiencies of this

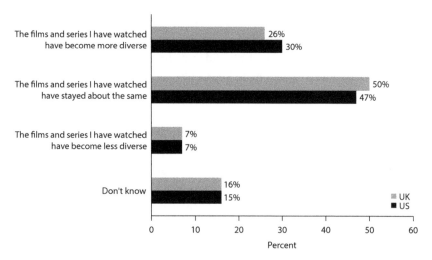

FIGURE 7. For the following question, by "diverse" we mean films or series offered in languages that you do not usually watch, made from a different time period, or outside of your usual/favorite genres. Overall, would you say that the films and series you have watched on Video on Demand platforms are more or less diverse compared to before you started using these platforms, or would you say they are about the same?

Unweighted base (for UK survey): All UK adults online who have used video on demand platforms to watch films or series (n = 1,692).

Unweighted base (for US survey): All US adults online who have used video on demand platforms to watch films or series (n = 1,042).

scenario: "I am not annoyed by that. I can keep watching the same in these genres because I'm interested in them. I'm not too interested, for example, in romantic comedies. So I will feel annoyed if they recommend me a romantic comedy." But this view was not typical among the sample. Indeed, many spontaneously bemoaned the perceived lack of variety among suggestions, complaining that the recommender system always presented the "same" content. Participant 4, for example, cited his annoyance that the system suggested "old films," his idiom for content he has already seen. Similarly, Participant 18 spoke of his efforts to whittle down his recommendations because he had seen most of them before: "I guess they're not always right [*laughter*]. I've seen 'suggested,' 'recommended for you,' 'featured TV series,' and they're not really exactly like, for example, a film I watch—I have watched already. So, yes, when they say, 'you watched this one, then here you go, ten films.' From these ten films, maybe five, I would watch five, and maybe three I would have already watched. But it is a good thing, a good tool, to see some stuff that you haven't seen and that you probably would like, so I use it." When asked whether she watches "more of the same" or a "broader variety" of different content, Participant 23 replied, "Yes, they just always send you to exactly the same sorts of shows, which I personally don't like because I'm not the sort of person that will watch exactly the same genre every single time. No, I haven't experienced that I actually focused just on those sorts of movies or TV shows. I find it's pretty difficult sometimes in Netflix to find something. I do feel that. I do feel like because I like *Friends*, then they just show me shows exactly the same as *Friends*."

Participant 12 acknowledged algorithmic suggestions on his interface but ascribed them limited influence because of his feeling that they circumscribe his choice rather than introduce him to exciting new content. He said: "but, for me, no, I don't care about the recommendations. Sometimes I search or ask friends. . . . I don't care about the matching. No, I want to see something outside of my comfort zone, viewing history or something, yes. Or something new." Later he continued: "For me, I don't like it, because it limits your choices sometimes. Because you want to discover something new and they just collect what you have watched and maybe something you like. I don't like it. I like to make it still open for everyone to see something new." This view emerged in a few interviews.

A significant subtext of the folk theories about filter bubbles and algorithmic recommender systems was the belief among some respondents that, ontologically, human recommendations provide more variety than those calculated by algorithms. Participant 3 said he trusts human recommendation more because of these perceived benefits surrounding variety: "The chances of discovery are bigger if I read a human recommendation. The way these recommendation systems work, it's like creating an echo chamber for you." Such respondents assume that word of mouth or critics' reviews are a priori less likely to confirm, rather than challenge, tastes. This belief is the folk-theory equivalent of scholars' mathematization-of-taste and filter-bubble arguments.

For Participant 19, who believed the recommender system was ineffective in limiting content to a satisfying choice, the sheer quantity of films and series on offer made her more likely to watch more of the same:

I end up watching more similar things now. Just because if I want to change to something I would have to actually—For me to feel like, "Okay, I'll make a decision to try something that is not . . ." it's out of my comfort zone, because it's not something that's been recommended. I will have to look into it and see, "Okay, let's see what it's about before trying it." Then you have to look at the trailer, it's true. You can take a long time just watching trailers and trailers and trailers, so I think it limits a bit more because you have to actively look for something different. I'm not sure if I do it as often now.

Later in the interview she reflected on how her viewing habits had changed since when she only had cable television.

I always used [to use] TV as background noise. I can be doing anything else and having the TV in the back; by doing that I would be up to date on what TV series were out and trying new stuff because it was just that.... Even though I wasn't paying attention, it was in the background, sometimes I'd think, "Actually, this looks interesting. I want to keep watching it." Whereas now, I would just browse to it and say—Sometimes there is too much content, and I don't know what to choose. Sometimes that makes me just stick with what I know, with old TV. Maybe not trying the new ones because there are so many shows coming on almost every month, it's very hard to keep up with all of them. That makes it more difficult for me, I think, to choose. It has changed to the fact that I now have to look for it. In many cases, I even . . . let's say I had cable TV, it probably would be different, it would have to be on all the time. I would have it in the background as well, and just keep watching all the shows.

These comments make a crucial ontological assumption: cable television broadcasters furnish variety; catch-up services and streaming portals do not. Thinking in technologically deterministic terms, this participant implies that reduced variety is the fate of a change in delivery system. Her comments also resonate with the research, presented earlier, that feelings of too much choice can lead to the consumer making no choice at all.

Participant 33, who, as previewed above, was perhaps the most forceful in rejecting recommender systems, was simultaneously among the most optimistic that VOD has not circumscribed the diversity of his film selection. Asked to explain, he said:

I think it just depends on the streaming platform, I guess. Because if you watch MUBI, you know you're going to get a more diverse kind of selection. If you watch Netflix, it's just the same thing all over and over and over and over again. Then it might have that one original series every now and then, but it doesn't really make that much of a difference. Sometimes you're like, "Oh yes, there's this foreign film." But I don't see that many foreign films on Netflix, they're mostly English-language films. There might be some exceptions, but most of the content feels like it's mainstream Hollywood, or Netflix own productions, and all those kinds of things. If I watch MUBI, I'm going to be like, "Yes, I know that; that's going to be all over the place," you know?

This participant, a sophisticated viewer with access to several VOD services besides his regular DVD, Blu-ray, and cinema habits, held the belief that Netflix's catalog and recommender system provided poor recommendations and selection options, "just the same thing all over and over and over and over again." Nevertheless, he posited that he was able to compensate and *self-diversify* his audiovisual diet on account of his multiple points of access to film, including MUBI and regular trips to the cinema and DVD shop. His way of speaking on this subject epitomized the folk belief in "free will": the savvy consumer has the potential to foil the machinations of media organizations. The user is ultimately the master in control of what he or she watches and must assume responsibility to transcend the filter bubble. The unspoken implication is that some may not have the access or knowledge or will to break out of the echo chamber.

A number of respondents conceived of their experience of consumption diversity in a similar way. When asked whether she thought since starting to watch VOD she viewed more of the same or a greater diversity of content, Participant 18 replied: "Hmmhmm, yes. But, even if that happens, I guess— Like, if you wanted to watch a different genre, for example, you can just go to the different genre and go through [them]." For her, the lazy habit to rely only on the homescreen recommendations was the root cause of any user's filter bubble. Overall, this respondent was ambivalent about the capacity of the tool: despite unsuitable suggestions that "don't understand" the user's taste, the recommender system remains useful for new discoveries and thus, in her opinion, contributes to diversity and can challenge preexisting tastes. Participant 22 proposed a similar view, suggesting the many ways that any given user could easily break out of a filter bubble:

> Well, I will say that I do watch other types of films also. It's not only I watch all the time thriller movies. That's why sometimes in the list, if everything is—the algorithm one that they will choose for me directly which ones would I like to watch. Instead of that, sometimes when I just want to watch a comedy, for example, then I go to "Genre" and select comedies. Then based on that I will choose whatever I would like to watch. Or still I can do the same searching things, so "best comedies of this last year" or something. Just narrow it by the time that I would like to watch something sometimes. Yes, I go directly and select the genre that I would like to watch mostly. So if I would

like to watch something about science fiction also . . . well, the fact that the algorithm will give you the same genre all the time, because you're watching the same thing all the time, probably yes, it will work like that. But if I try to mix it, like yesterday I was watching a science fiction movie, tomorrow I will watch a horror movie, then it will give different options. That's what I've seen on my list too. It's not only narrowed to thriller movies. It also gives me options for comedy movies.

For this viewer—who seemed to assume less algorithmic personalization to the genre rows than other parts of the interface—platforms like Netflix might not provide a full palette of content variety at first glance. And yet, in her opinion, the ultimate culpability for any filter-bubble problem rests squarely on users. Unengaged viewers who do not provide the algorithm with the full diversity of their taste are to blame for any lack of recommendation variety or effectiveness.

The contrasting interview responses presented in this subsection suggest how we might reconcile the seeming paradox between, on the one hand, the survey results that indicate most say their viewing has become more diverse or has remained roughly the same and, on the other, the chief complaint that selection on streaming platforms is inadequate. First, among those who feel that VOD recommender systems offer more of the same, there seems to be one exceptional way of speaking: sameness is actually beneficial. This discourse is generally consumerist and utilitarian, for example, Participant 4's satisfaction that he does not have to wade through rom-coms on Netflix because he would not watch them in any event. Second and much more prominently, another folk theory recognizes the problem but insists users must surmount it by themselves. Participant 33 is exemplary in this vein, believing in the "free will" of the user to break out of the filter bubble and compensate by engaging with film and series culture beyond Netflix. Third and finally, there is an attitude that bemoans the filter-bubble problem and believes there is essentially nothing to be done about it. These respondents posit themselves as subject to the whims of large nontransparent corporations and forces beyond their control. According to them, "times have changed" since the days of cable television and video shops, either because of the ontological qualities of algorithmic recommendations or because of morally suspect companies' implementations of this technology. The last two ways of speaking, then, hypothesize a radically different approach to film and media culture and the filter-bubble issue: one in which the user has the ultimate free will, and the other, often more nostalgic assessment, which posits that in the new-media environment, the user is bound to follow the fate that business or science determines.

IV. TYPOLOGIES AND THEMES

The surveys and interviews offer incredibly rich sources for understanding how viewers select films and negotiate VOD recommender systems among a host of other influences. For reasons of space and coherence, I cannot explore all of that information in depth here. For example, why do women—in both the UK and US surveys—report using critics' reviews and also aggregators such as Rotten Tomatoes at significantly less rates than men? Does it have anything to do with feelings of a lack of female representation among professional pundits, a prominent discourse in the news and social media?[23] Why do the young and US Hispanics trust critics much less and recommender systems comparatively more? These questions will no doubt inform my research elsewhere for years to come.

In the final section of this chapter I would like to instead deliberate on a few distinct categories of users, and in particular their choice motivations and repertoires. One crucial methodological distinction bears emphasis here. The previous three sections responded to set research questions with quantitatively assessed survey responses, which I then triangulated with rich qualitative data. In contrast, this fourth section emerged inductively from the interviews alone—from listening, coding, and categorizing self-descriptions of participants' behaviors, language, and beliefs. The following discussion suggests that frequency of use (i.e., how many films and series any given user consumes) may be just as if not more predictive of repertoires of using recommender systems and other information sources to select films and series than traditional demographic labels. It also details how the *importance* users ascribe to the role of film and series in their lives may be most predictive of all.

Information Limiters and Information Gluttons

The interviews revealed a number of distinct user typologies, distinctive routine behaviors of consulting information for selecting films and series. One typology I call Information Limiters. Several of the interview participants (specifically: #5, #7, #16, #17, #25, #27, #32, #33, and #34) demonstrated this type of behavior: purposely limiting the information they consumed about a certain film or series. Their objections often involved seeking to avoid "spoilers," but also, in more positive formulations, aiming to build anticipation and "adventure." Participant 25 was perhaps most extreme in this regard: "I don't watch any trailers for films. I actively avoid them, so if they come on, I will put the TV on mute and cover my eyes, which sounds ridiculous. . . . I actually leave the cinema when film trailers come on to avoid them." Several noted that this repertoire applied more for some viewing situations than others. For some, limiting information was more relevant for low-stakes content. One participant (#8), for example, noted that when seeking "light entertainment" or when she has preexisting knowledge (in particular: she has seen the prequel), she does not feel the need to acquire further information. Other participants indicated the opposite tendency: content for which they maintained high expectations as a meaningful experience would lead them to purposely limit exposure to further information so as to preserve an element of anticipation and surprise. Participant 14 said that for her favorite films (Marvel Universe), tips or other foreknowledge were superfluous and unwanted: the release date alone was sufficient. Participant 32 spoke about limiting information for her favorite directors. She may watch the trailer but will otherwise eschew news and other sources in order to avoid anyone "shap[ing her] opinion." In contrast, unknown or lower-rated filmmakers or content required closer scrutiny so as to make a viewing decision. In discussing her decision to see *Us* (2019), she intimated that the director (Jordan Peele) functioned as an imprimatur of quality; she needed neither trailers nor any other information to evaluate potential opportunity costs.

On the other end of the spectrum, interviews revealed a number of what I call Information Gluttons. Some users, particularly those of the cinephile or series junkie type (e.g., #31), were more or less constantly consuming information about upcoming content on blogs, in magazines and newspapers, and via a wide diversity of other sources such as social media, ads, reviews, interviews, and production news. They often used language like "I heard about the buzz" or "hype" and found it difficult to pinpoint precisely where they first noticed any particular content because of the prodigious amounts of information about films and series they consumed on a daily basis. Participant 6, to cite another example, spoke of the hours of research he sometimes executed before deciding to watch a film, including reading dozens of user reviews, especially of films with which he was unfamiliar from other sources. Participant 11 was particularly interesting in this regard. He detailed conducting significant amounts of research but then frequently not following through to watch the content; his threshold to actually consume a film or series itself is particularly high. This act of information seeking—in his case triggers from ads or word of mouth from friends and colleagues, Facebook, or trailers on YouTube or IMDb—becomes almost an end to itself and satisfies his curiosity regarding the content: "Most of the time the next step is I don't end up watching that movie." Information-seeking activities displace and even replace his need to watch the film or series.

Lazy Choosers

Another typology among the interviewees I term Lazy Choosers. Key examples among the sample were Participants 1, 7, 21, and 28. Lazy Choosers do not undertake significant effort to inform themselves about content; they do not typically read critics' reviews or articles about films or series, nor do they consult online user forums. Instead, they primarily take word-of-mouth tips from "experts" in their immediate social circle: for Participant 1, his brother; for Participant 21, his friends and flatmate; for Participant 7, two friends who are critics back in her home country, with whom she often engages via social media. When selecting content on VOD platforms, Lazy Choosers make little effort to read synopses or spend much time browsing, but rather click on titles they have "heard of" among the curated range of recommendations. This is usually a tip from their social circle or Facebook newsfeed, but it could also be foreknowledge of an adapted intellectual property. For example, Participant 7 reported watching *Tidying Up with Marie Kondo* (2019) because she had heard of Kondo's books. In this sense, the presentation of a recommender system forms the second part of a triage: the location where a previous "trigger" is recalled. Users of this type tend not to trust Netflix in the abstract; nevertheless, the interface arrangement provides a field to apply prior information from other sources.

Typically possessing a low finishing imperative, Lazy Choosers often do not complete the content they embark upon watching. Rather, they switch to something new once they become bored, which happens frequently. (For example, #21 reported giving up on most series after the first episode, especially if it was not recommended to him by one of his friends.) Some examples of this type showed a low understanding of their own taste— or, more specifically, an inability to predict enjoyment based on primary information sources such as title, image/poster, and description, perhaps because of their overall low engagement with film and series culture.

Although he spoke about receiving good tips from his friends and his flatmate, Participant 21 said that when browsing on Netflix he had serious difficulties finding content that could sustain his attention. Scrolling without end and trying out content—but unwilling to invest the time and effort in finding out more about the programming (he reported that occasionally when he hears about a show, he may at most google it for a "rating")—he will usually end up disappointed.

Lazy Choosers also tend to prefer programming (strongly episodic series without series arcs) that does not require much attention because they use VOD as a sort of "wallpaper" while multitasking. For example, Participant 1 spends much of his viewing time on his phone, while Participant 7 reported commonly using VOD as a background accompaniment for eating, sleeping, or sex.[24] (These sorts of users—#1, #8, #21—often said they do not like going to the cinema, and do so rarely, because they feel restless not being able to multitask by using their phone or speaking to friends. In equal and opposite measure, #28 *preferred* cinema because while watching films at home she is easily distracted and likely to quit watching to complete a household chore.) Members of this group eschew paying for their own account. Rather, they piggyback on someone else's password (#1, #7) and utilize free-trial periods. This behavior is indicative of their overall high attention to cost-benefit calculations and the relatively low value audiovisual content holds in their lives.

Some of the Lazy Choosers across the sample were, by definition, in terms of hours streamed, heavy users. Nevertheless, their engagement with the content was overall very low, because of their frequent multitasking. For this reason it is helpful to create a category beyond heavy users and light users: high-stakes and low-stakes users. Lazy Choosers are low-stakes users, for whom film and series culture is not a primary interest. Participant 13, a low-stakes light user who usually passively relies on word-of-mouth tips to discover content, exemplified this type's thinking in his interview:

> I don't know if I said this before, but I'm not a big fan of watching movies, so I didn't go on a big search about it. But we live in a society where people talk about different subjects and talk about movies or TV shows, it's like [a] big [deal], that's how they talk about some movies they watch or, recently, the best movie they watched in the last few weeks or months. Then I pay attention, but it never comes from me, it comes from them. And I ask them something—if it's good, I ask them about what the story—what the movie is about then.

For low-stakes users, films and series play an insignificant role in and of themselves. As we shall see below, audiovisual content is often more important as a means to stimulate conversation with friends than for personal enjoyment during viewing. In contrast, high-stakes users simply care more about films and series, which maintain a relatively prominent position in their hobbies, interests, and personal sense of identity and self-worth.

"Keeping Up": Heavy Users vs. Light Users, High-Stakes vs. Low-Stakes Users

As mentioned, the quantitative results illuminated a few areas where traditional demographic categories such as age, gender, race, or class yielded differences of statistical significance. Nevertheless, the interviews indicate that other categories may be at least as compelling if not more so: the differences in behaviors and understandings of VOD recommender systems (in the context of other sources and influences) between heavy users and light users and, more precisely, between high-stakes and low-stakes users. These results both partly confirm and partly complexify earlier audience studies.[25]

Heavy, high-stakes users tended to speak about different search-and-browsing protocols. For example, Participant 34, a heavy (daily) VOD user and relatively high-stakes user overall, suggested that she only ever looks at the Recently Added category on Netflix because she had "already gone through genres so many times, I know every single film, so what would I do? I just go into the Recently Added." However, a few times a year (e.g., after going on holiday), "if I fall behind, if I'm not really up to date, then I will go to the genre search."

Heavy, high-stakes users also maintain distinctive features in relation to word of mouth. Although word of mouth tended to be by far the most important information source both in the surveys and the interview sample, heavy high-stakes users (e.g., #4, #22, #31) often reported tips from friends, family, and colleagues as a less important source. Oftentimes, this was because these users were frequently functioning as *providers* of word of mouth themselves. These WOM Providers, e.g., #17 and #31, reported blogging, writing amateur reviews, retweeting critics' opinions, or Facebook-posting recommendations. In turn, any of these heavy, high-stakes users tended to eschew aimless browsing (cf. a low-stakes user like #21 above) and were often goal-orientated searchers on VOD, because they were also Information Gluttons and thus had extensive foreknowledge of film and series reputation. (Indeed, these users seem to overlap with "superpredictors," a small subset of humans who, studies have shown, consistently, substantially, and easily outperform algorithmic systems in recommendation accuracy.)[26]

The heavy, high-stakes users among the interviewees tended to be confident in their knowledge, opinions, and taste. Many were skeptical of certain sources, speaking of them in disdainful and sometimes hostile terms. For some this constituted derision of user forums; for others, critics or Rotten Tomatoes. For example, Participant 31 never uses Rotten Tomatoes because of her confidence in own taste and her feeling that aggregated critic and user tastes do not reflect her own opinion: "I don't really agree with their scores at all. . . . A lot of my favorite films have terrible scores." This sentiment was echoed by others (e.g., #27, #30, #33). Participant 34 held the opposite outlook, suggesting that user forums and Rotten Tomatoes are most important for her, whereas, in her view, critics often yielded poor recommendations because the latter did not share her opinions. But she too was equally confident about her taste and the sources she needs to consult (and to avoid) to find suitable content. These heavy, high-stakes users and WOM Providers tend to believe more in the "free will" of the user, rather than folk theories about algorithms and media company machinations dictating content. These insights confirm, first, prior theoretical research into "consumption capital": the more a consumer invests in learning about culture (watching many films and series, reading reviews, surveying interviews and production histories), the more he or she develops a more fine-grained set of preferences and the ability to make media choices that satisfy that taste.[27] Second, these participant views align with earlier empirical research that indicates how heavy users remain less susceptible to persuasion (e.g., from advertisements) because of their knowledge, confidence, and "rigid preference structures."[28]

The interviews—and the general differences between low-stakes and high-stakes users—yielded fascinating insights about content choice in the age of Netflix and about the general functioning of film and series culture. Inadvertently, the results of the study contradicted the neoclassical economic rational agent-consumer model, whereby any given user simply attempts to maximize utility by informing him- or herself about the film or series and then selecting wholly on the basis of prior viewing history or within a narrow portfolio of favorite genres. Rather, a social interaction that I coded under the rubric of *Keeping Up* was much more important. This topic abounded across the interviews and various interviewee types, proving to be one of the most prevalent, durable, and enlightening themes.

Keeping Up expressed itself in several ways among the various subgroups. For some (mostly the heavy, high-stakes user category), this topic denoted staying up-to-date with cinema and television culture: for example, feeling compelled to watch tentpole movies or prominent series. Among some there existed a more socially discursive notion: keeping up with critics' opinions about new films or series, or with social-media discussions. Participant 6 spoke of his routine to "browse the website every week" to find new films from his favorite national industry, Bollywood. Participant 33, a student with professional plans to enter the media industries, spoke about watching new films almost at random to keep up with cutting-edge trends in stunt work and cinematography. He also spoke about keeping up with the films recommended by critics he respected, thus staying current on idioms of taste. He described watching films in order to test and compare his opinions against critics and also cinephile friends, with whom he often had post-screening discussions. The act of sharing and speaking about films—including disagreeing—was both pleasurable and central to his media experience.

This example gestures toward the most common meaning of Keeping Up across the interview sample: keeping up socially among friends (and sometimes family). Films and series' role as important topics in social situations remained one of the most powerful aspects of interviewees' reported media use. Keeping Up transcended individual sources, although users most often spoke about word of mouth, social media, user reviews, and critics' write-ups in this vein. But even VOD platforms and recommender systems were sometimes discussed for their social aspects and elements. For example, some mentioned VOD-internal user ratings or the Netflix categories Trending Now and Popular as providing a feeling of sociality and being able to see what others are watching and enjoying. For Participant 19, films and series feature as a key topic of conversation at work and she counted her colleagues and boss as her key word-of-mouth influencers. "We have similar interests. The way they perceive things, perceive information, are quite what I like. I follow their recommendations a lot," she said, mentioning *Lucifer* (2016–2021), *Daredevil* (2015–2018), and *Arrow* (2012–2020) as series she learned about through—and later chatted about with—her coworkers. In general, she closely attends to what others watch and like, eliciting tips via social media (in particular Facebook), user ratings on IMDb and Rotten Tomatoes, and the Trending Now section of Netflix, which provides a glimpse of what is popular at the moment: "I tend to go to Trending and see what's common. I think one of the reasons I really like to watch films and series would be because then you can talk about that with someone else that is probably watching it. If someone says, 'Who is watching?' 'Let's watch it, then we'll talk about it later.' It's just like a discussion topic."

Participant 21 was also interesting in this regard. A low-stakes Lazy Chooser ("shows and movies are not a huge part of my life"), he is nonetheless a very socially driven user and maintains an active social life and routines. He meets friends twice per week at a pub, where films and series feature as one of the most frequent items of conversation among the group. Audiovisual content is a way to share information and socialize with his mates, who remain much more devoted fans than he. Participant 21 spoke about watching films and series as a way to keep up socially, to be able to participate in discussions and express his tastes among his in-group. Several interviewees spoke in this vein. For instance, Participant 1 reported wanting to see "what people are talking about"; others emphasized the importance of event films and series, "that everyone has to see, like the new *Star Wars* or something" (#21). A few participants highlighted social media as an important conduit for keeping up: Participant 24, a light film and series viewer, said he would "take a second look" and potentially watch content if he read many references to it online. Overall, his discussion of films and series revolved around following trends and wanting to experience "what people are talking about." This behavior corresponds to cultural theories about the pleasures of shared tastes, blockbuster movies and series and books that are consumed and enjoyed precisely on account of the scale of their cultural reach and their common social currency: watercooler moments.[29]

Participant 34 displayed a variation on this theme, speaking of watching films and series recommended by word of mouth as a way to *understand* her friends better: "It's just when you're talking to your friends or what films that you watch or any TV series and they tell me about which series are good. I know I can trust their taste. It's just pretty solid. Because it's their interest and I like them, so I want to share the same experience, these two things." Word-of-mouth tips have led her to reflect on her friendship and to learn more about this particular friend and his or her taste—added value to the recommendations themselves and the basis for subsequent discussions and conversations with the recommending friend.

Many participants (e.g., #2, #6, #8, #9, #11, #12, #13, #14, #16, #17, #18, #19, #21, #28, #33) spontaneously reported that they would watch films and series outside of their preferred genres when invited by a friend, colleague, or family member or when the "leader" among their circle of friends had decided on a particular film (or, less often, series binge). A few (e.g., #14 and #18) used the term "group pressure" or "peer pressure" to describe these situations, and some said they occasionally drew a line: Participants 14 and 19 said they declined for horror films or if there was an actor (e.g., for, #19, Jim Carrey) whom they particularly disliked. But almost all the others reflected on such experiences in a more positive manner. Participant 17 perhaps summed up this sentiment most succinctly: "To be honest, within reason I'm pretty up to see anything. The company is more important to me than the film really, even if it's a really rubbish film. I saw *Mortdecai* [2015] with Johnny Depp years ago with my friends, and it was utterly, absolutely the worst thing I'd ever seen at the time. But we still talk about that now."

Indeed, word-of-mouth tips and the potential for a positive social experience in general seemed to trump personal genre tastes. Participant 8 offered the following response when asked about her favorite and most disliked genres: "I wouldn't watch anything horror. I'm not really into crime dramas either, things like—Is it *Silent Witness* [1996–present]? I wouldn't watch anything like that. But when *Bodyguard* [2018] was on, because everybody was talking about it, I got into that. Which isn't something I would normally go for, but it was good." Others, such as Participant 11, spoke about this topic in a nearly identical way: "I don't think I would ever decide to watch a horror movie that, you know—yes, I would never decide to watch a horror movie, but in some exceptional cases, for example I have friends who are like massive fans of horror movies and just to spend time with them if they decide to watch a horror movie I wouldn't hesitate to—I wouldn't go out of my way to not watch it." These statements resound with other empirical revisions of genre theory. Many single-subject studies of *Star Trek* cosplay or other devotees of a solitary franchise or storyworld unintentionally caricaturize "fans" who supposedly maintain one favorite type of film or series from which they never depart. And yet research shows that most people consume different genres promiscuously.[30] The social pleasures of films and series—with whom you spend time before, during, and after consumption—may be just as important, if not more.

Hundreds of op-eds and academic articles have diagnosed a newly atomized media sphere in the wake of VOD and recommender systems. And yet the above comments suggest that the crucial social functions of film and series culture, so integral to pre-streaming generations, may still obtain. Scholars such as Richard Maltby, Robert C. Allen, and Annette Kuhn have demonstrated that twentieth-century viewers saw cinemas as "sites of social and cultural significance" having much greater significance than "what happens in the evanescent encounter between an individual audience member and a film print."[31] Indeed, according to Kuhn's empirical investigations, for most audiences, "memories of cinema have revolved far more around the social act of cinemagoing than around the films they saw."[32] The evidence presented in this chapter suggests we can extend such claims beyond physical cinemas and their attendant social pleasures of sharing popcorn and the post-film beer; discussions of Netflix series in pubs, via text or social media, continue this important purpose of watching, somehow, *together*. The fact that today most people consume films and series alone or in small domestic groups does not mean that the overall experience of film and series culture has become individualized or divorced from social purposes and needs. Such comments also further suggest, looking forward, that no matter how targeted algorithmic recommender systems become, such tips will most likely never replace the social pleasures of word of mouth. For many, the film or series itself is only part of a larger equation: social exchange. An ongoing British Film Institute (BFI) longitudinal national survey, with fieldwork conducted both before and during the COVID-19 lockdowns, bears out this finding. Attitude statements that suggested the value of film as a social currency, such as "I like to recommend films to people I know" (63%) and "The cinema is an important contributor to culture in its local community" (75%), attracted the highest levels of agreement.[33]

Humans vs. Algorithms

These essential social aspects of film and series consumption return us to fundamental questions surrounding VOD recommender systems: whether these tools are reconfiguring culture and replacing humanistic functions and functionaries such as film critics, whether a new "algorithmic culture" is superseding human recommendations. In anticipation of the afterword to this book, let us begin to examine this question by considering how the interviewees spoke about human-generated suggestions (particularly from critics) versus algorithmic recommender systems.

As stated, and shown in Figure 8, the survey demonstrated large majorities who said, if forced to choose, they would trust human critics (UK: 74%; US: 64%) over computer algorithms (UK: 7%; US: 12%) for a film or series recommendation, with a substantial minority reporting they did not know how they felt about the topic (UK: 19%; US: 24%). In the survey design, this question figured as the respondents' final task and was always only meant as

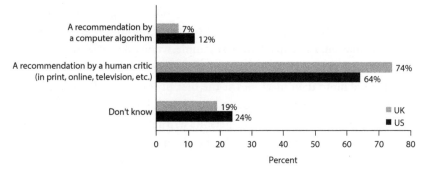

FIGURE 8. Now thinking about watching a film or series on ANY platform (i.e., NOT limited to Video on Demand) ... If you had to choose, which type of recommendation are you more likely to trust? (Please select the option that BEST applies.)

Unweighted base (for UK result): All UK adults (n = 2,123)

Unweighted base (for US result): All US adults (n = 1,300)

a blunt, nationally representative indication of feelings on the subject. The individual interviews were much more illuminating. They indicated that real users' opinions may be much more mixed (and contradictory) than the survey results suggest. For instance, answering the question one way was not dispositive for larger behavioral patterns. Participant 34 was one of several interviewees who responded that she "definitely" would trust the recommendation of a human critic over an algorithm. However, other parts of the interview intimated a more complicated picture: she said she had long since stopped trusting critics' reviews, as she felt that they did not share her taste. She reported negotiating VOD recommendations with user reviews on IMDb, Rotten Tomatoes, and other online venues. Participant 31, who possessed a uniquely strong credibility in both critics and VOD recommender systems, found the question too difficult to answer because the two sources maintained different functions in her experience of film and series culture: "That's a really hard one. I mean, as well, there's taking into consideration things like Reddit, as well, and they, kind of, talk about a lot of films, just loads of different people with their opinions, and I think that's really good. I think film should always be talked about like that, by critics and just general audience. I honestly think both have their merits. I don't think I could pick one, definitely." For this participant at least, VOD recommender systems cannot conceivably replace critics, if only because the two sources serve radically incommensurate roles in her media life.

As stated, across the sample of interviewees, responding to several different questions, many took a strongly humanistic stance toward the general issue of recommendation. Participant 32, for example, when asked which she trusted more, immediately said "human critic," explaining, "I don't know. I kind of trust people more than machines at the best of times, and I would like the social aspect in a way. Even if it's someone I don't know, if I'm just reading a lot of articles, you get a feel for the person's writing style and personality, and that's just more fun." In this example, the social engagement of an imagined dialogue with a human critic represents a more important characteristic than the "accuracy" of the recommendation. These responses gesture to the many functions of criticism: reviews help us learn about new cultural products and inform consumption decisions, yes, but they also represent a social mobility strategy, a means of testing taste against others and of illuminating cultural history.[34]

Many interviewees stated that they would trust a human critic over a computer algorithm. Nevertheless, very few could actually name a critic whom they read or watched on a regular or even occasional basis. Of thirty-four participants, only a fraction said they ever read critics. Of these, only a handful said they had a regular or favorite critic. Five respondents mentioned Mark Kermode, reviewer for BBC television and radio as well as the *Observer* and thus Britain's most visible film critic; one mentioned Xan Brooks of the *Guardian*; and two others mentioned "a Spanish critic" or "a critic from Saudi Arabia," referring to their respective countries of birth. The fifty-four-year-old Participant 27 reminisced about Barry Norman's 1980s BBC1 film review show as well as *Halliwell's Film Guide*, the British equivalent of the Leonard Maltin capsule-review compendia. With such responses in mind, we may need to think of the survey credibility differentials of forty to fifty-five percentage points less as a strong endorsement of critics, than as an indicator of exceptionally low trust in recommender systems.

In general, even among those who said they engaged with critics, responses were more ambivalent than the stark quantitative results on this question might suggest. Participant 23, for instance, said she triangulates recommendation sources because of bad past experiences and for cost-benefit reasons. When asked whether critics' reviews were important to her, she replied, "Oh yes, totally. Yes, I would prefer the recommendation of a critic rather than, I don't know, someone on Facebook saying that something is good. I do trust more the critic." But when asked whether there were any particular critics she trusted or followed, she said: "No. To be fair, no. I sometimes tend to read all the internet reviews. For example, there are particular websites that will review the movie and will tell you what their opinion is. I will read those, but I don't have one specific one I go to apart from IMDb. I know it's [different] with many different people."

Participant 22 also maintained an ambivalent attitude toward critics, highlighting the fact that her trust in them and use of reviews fluctuated depending on the type of film she wanted to watch and her idea about whether critics might share her taste. She associated critics with "artistic films" and "Oscar award–winning movies" and eschewed them especially when she wanted an easy-watching experience.

> That's why if I just want to [see] something that I really want to enjoy and chill and have a good day, a cozy night watching a movie, I will go for what the audience say. For example, people will say *Iron Man* [2008] is a bad movie but people enjoy it. That's why I try to focus more on the audience score rather than Rotten Tomatoes. But if it's only got, say, one out of five stars on Rotten Tomatoes, then I think I also use that as a baseline for me to choose something better.... Sometimes it's also about just watching a film in a relaxed way and not having to pay attention to all the artistic qualities of the film.

For these situations, she tended to rely on IMDb scores, Google's percentage of how many users enjoyed the film, and occasionally Rotten Tomatoes. These, and especially Google's percentage, appealed to her because "the people [are a] little bit less, I think, [judgmental]." For Participant 22, the incongruence between her taste and critics' opinions—especially involving films from her native country—inspired anger and annoyance, so much so that she tended to avoid reading them:

> I do think that critics are interesting and are very important for me to select something. At the same time, when I've been reading some critics from some movies that I enjoyed a lot I just feel that they don't get the— especially when they're Mexican movies. They don't get the essence that it has to be implemented on the movie and, because it's a foreigner critic, person, probably I don't enjoy it that much. I do rely on them, and I think it's a very important job for them to do, but I think it also depends on what I like. If it is something that I will feel probably a little angry—not angry, but sad—over the comments that they're saying about a movie that I liked then I wouldn't listen to them. Or I wouldn't read what they're trying to say about this movie.

For Participant 22, then, even though she did regularly engage with critics' opinions, whether in review form or via aggregators like Rotten Tomatoes, her engagement was selective (based on the type of content she was seeking out) and often fraught with negative emotions. Overall, her suggestion that critics' reviews were only applicable, necessary, or desirable for certain viewing situations parallels, somewhat in reverse, interviewees' opinions on VOD recommender systems, which feature as a primary source in "easy watching" situations.

In the interviews, respondents often used the question of whether they trust human critics or algorithms more as a way to pivot to the general topic of credibility. *Trust* became a term most often explicitly invoked in relation to human word of mouth (and critics for those few who admitted reading or watching reviews). But others foregrounded their specific trust in user ratings (e.g., #6, #10, #34), parents' reviews (for #9, the mother of a fifteen-year-old youth), or other specific sources. Participant 24, a light user, cited time and cost-benefit reasons for why he consults user ratings over critics' reviews: the former are superior because he can consume them more quickly and, in his calculation, they remain more reliable for determining content quality. According to Participant 34, despite having subscribed to cinephile movie magazines as a teenager and young woman, she had become gradually disenchanted with critics' opinions, especially as they concerned her own taste in films and series: "I trust more viewers actually because when I do [read reviews], then I often disagree with the experts. I think I developed that kind of habit." Several participants volunteered that they do not engage with reviews because pundits' evaluations failed to correspond to their taste. Participant 15 explained that critics are not important because they do not match her taste and are too "serious" and "biased" to appreciate entertainment value: "Probably I feel like normal people, you and I, can relate; they don't necessarily—a movie is something you just want to enjoy, you don't need to make it very serious. So, I feel if every Tom, Dick, and Harry gives their opinion, I'll value it more than a film critic, who may be a bit more biased. They may be too serious to look at the fun parts of it. That's what I think." This was the most common theme. Participant 28, an older, strongly socially oriented light user, explained that she does not consume criticism because "I think, well, everybody is different, and what one critic might say is good I'll most probably think, 'What on earth are you going on about?' You know, it might not be of interest to me." Most arresting was Participant 10's abrasive reckoning with critics: "No, I don't really trust those because it's just people. Anyone could do that." This type of statement applied more commonly to user ratings.

Indeed, a few participants (e.g., #20, #27) deployed almost exactly the same language to rationalize why they trusted critics but refused to engage with lay user ratings and reviews: "I'm not really going to take the views of unknown strangers. You don't know the agenda behind some of these things anyway. It could be a rival company.... I don't trust them, no" (#27). It is noteworthy that, especially among older users who rarely if ever use VOD, a similar discourse prevailed but substituted other concerns for user reviews or algorithms. Participant 29 said he doesn't "use social media that much, but I'm not sure how much, even if I did, that I'd be influenced by them, because you don't know who's writing it a lot of the time." Participant 30, a sixty-two-year-old avid reader of criticism, repeatedly and suspiciously denounced "marketing" and media companies' "focus groups," saying "I don't want to feel that I've been manipulated" by commercial imperatives into watching a certain film over another.

Particularly striking is Participant 33's formulation to express his trust in critics and distrust in "regular users." When asked why he did not read user postings on film sites such as IMDb and blogs, Participant 33 replied: "Because I don't trust those people's opinion, and why would I trust some random guy's opinion on the internet?" When probed as to why he reported not using Rotten Tomatoes either, he said:

> Because I don't trust them.... Yes. It's really simple, I really don't trust a lot of people's opinions on films, yes.... Especially because there's films that I absolutely love, and I was like, I have no idea how they got such a low score. It's probably because the mainstream taste doesn't like that kind of film. Like one of my favorite films that I watched in 2015, I think it was 2015, was *The Neon Demon* [2016]. I really loved that film, I loved it so much, and I know that it has horrible reviews, people hate that film. Then you have other films, and then you have the typical mainstream films, and you have all the *Avenger* films, and the *Avatar* [2009] and whatever, and they're like nine out of ten, and you're just like, please.

This statement echoes—in reverse—Participant 10's statements about critics and Participant 22's emotional response that reviews do not reflect her tastes and experiences.

This structural similarity deserves pursuing further. At stake in all of the comments about trust in humans over algorithms—and the credibility of certain types of human recommendations over others—is, first of all, a feeling of not wanting to be "pushed" into a media selection by an information source that the user feels is inauthentic, untrustworthy, or illegitimate. Even as they rely on sources to manage choice, people want to maintain a sense of agency.

Second, there is a clear need to have a feeling of belonging (or not belonging) to a certain taste community and the desire to have one's taste reflected and confirmed by a source that the respondent feels is authoritative. Several respondents suggested that they preferred critics' (or users') reviews that *confirmed* their tastes; they correspondingly disliked (and often avoided) users or critics who they believed did not share their tastes. This is striking because many of the filter-bubble arguments depend on the principle that "algorithmic culture" is unprecedented for the way it encourages users to confirm, rather than challenge their own tastes, and assume that this feature is ontologically related to the new recommendation technology. The current study suggests, however, that those users seeking taste confirmation do so likewise with human recommendations. The same principle obtains regardless of whether it refers to the traditional authority of the learned critic or the "wisdom of crowds" variety of user ratings and comments.[35] Overall, many people want their tastes affirmed and mirrored back to them.

Third, we should not discount the important social aspects of recommendation, even in a media world where consumers are trending toward watching content alone. Users see themselves in an imagined taste dialogue with a critic, a certain fan subgroup, a random user, or with the "people" writ large. These are perhaps transactional and fleeting dalliances, and not the years-long relationships that many have surmised to have transpired between readers and "star critics" such as Pauline Kael. But they serve an important purpose nonetheless. Indeed, it begs the question of whether VOD recommender systems, no matter how "accurate" they may become in future, could ever assume significant levels of trust, let alone replace critics or word of mouth. Such systems presently lack substantial dialogical and Keeping Up features and functions.

Fourth and finally, the question of "human vs. algorithm" is less straightforward than the quantitative results may initially suggest. To be sure, the respondents often spoke passionately and at great length about which sort of human recommendation they trusted more, and when asked directly nearly all said they trusted humans over algorithms. Nevertheless, this bare result should not imply that participants' interactions with human recommendations were always positive, or even that their experience with human recommendations was much more positive than with VOD recommendations; the large differential in trust between human critics and algorithmic recommender systems is perhaps more a reflection of the latter's exceptionally low credibility rather than a vote of confidence in the former. This single result also masks the extent to which respondents valued VOD recommender systems for certain fare and viewing situations (above all "easy watching"). The afterword, to which I now turn, reflects further on how the folk understandings explored in this chapter pertain to larger debates on the status of humanistic culture in the age of proliferating algorithms.

Afterword

ROBOT CRITICS VS. HUMAN EXPERTS

IN CURIOUS WAYS, users' folk beliefs about their fate or free will in relation to recommender systems broadly reflect the dreams and nightmares of
tech talk and scholarly critiques. Seen from a panoramic view, the debates
surrounding these and other new-media developments hinge precisely on
differing views of human nature and agency, on the one hand, and on opposing moral assessments of technological forms, on the other.

The techno-optimists in the industry, press, and academia share three
key assumptions. First and foremost, they (fore)see benign, innovative
technology. Transposing the promotional rhetoric of tech giants such as
Facebook into academic prose, these theorists assert that the very shapes
and communication models of internet-enabled networking are *per forza*
forces for good.[1] Second, the techno-optimists assume active audiences with
preformed tastes: unique human beings' exogenous, unique sensibilities.
They mostly ignore economic and regulatory frameworks; on the whole, they
neglect the effects that underlying content catalogs, the specific quantities and
types of available films and series, exert. To them, the idea of "personalization"
represents the fulfillment of a long-held goal: "democratic" access to, and
interaction with, media content. Third and finally, these commentators
often imply that consumers are now better connected among like-minded
communities of interest, a development they register as decidedly positive.

Techno-pessimists, filter-bubble theorists, and other academic Netflix
critics consider these developments in largely equal and opposite terms. For
them, first, this technology is pioneering but essentially malevolent, because
of its creators' and owners' corporate interests, but also because of the invisibility and impenetrability of its mechanics and processes. Lacking the access
or specialist knowledge to examine Netflix's or Amazon's recommendation

algorithms, many of these commentators default to worst-possible-case scenarios. Second, they see an illusion of choice, and certainly no true media pluralism, despite new proliferations of various media products. Users, they reason, are pushed toward outcomes by said technologies, business profit motives, or regulatory frameworks. For one of these commentators, for example, Netflix's recommender system "functions less as an expert" and more as "censorship in disguise."[2] In these assessments, with few exceptions, audiences feature as passive subjects, assumed to have more or less malleable tastes until acted upon by media companies. Where the tech-friendly commentators envision empowered "participants," techno-pessimists see vulnerable consumers at the mercy of essentially evil gatekeepers. Third and finally, these developments are polarizing (or have already polarized) society into like-minded taste siloes, gated communities of interest. VOD recommender systems, these interlocutors argue, fragment viewers into niches, disinhibiting common culture and fracturing the public sphere. Because viewers have more systems and channels of media delivery from which to select, and because they can increasingly determine the wheres and whens of media use, tastes are "splintering."[3] Unlike the techno-optimists, whose thoughts on this subject overlap, techno-pessimists code this development as unfortunate, dangerous, or narcissistic.

Despite their apparently divergent viewpoints, both sides seem to agree on three principles. First, that these developments are historically unprecedented, necessitating celebration and goldrush-entrepreneurialism, on the one hand, and hypervigilance and state regulation, on the other. Second, both sides see the need for new ways of thinking, which in practice often means discounting legacy media forms and information sources. Third and finally, both views make techno-deterministic leaps in their arguments. New technologies are the motor of cultural change, they assume; economics, regulation, and even human agency can merely respond to and at best slow or hasten these unparalleled and now inevitable developments.

Both views, the evidence presented in this book has demonstrated, are insufficient. The difference between the network utopianism of a Yochai Benkler and the surveillance-and-segmentation scenarios of Eli Pariser et alia can resemble the choice between a naive or paranoid view of human nature. These diagnoses depend, in crucial ways, on assuming either the best or worst intentions, respectively, of those who create, control, and consume media. These two otherwise incompatible worldviews both stipulate all-powerful technology and users acting in uniform ways. The echoes of these paradigms in folk beliefs—that "you can just go to the different genre" or "there is no way that you can run away from algorithms"—are evident.

LITTLE TRUSTED, LITTLE USED

The preceding chapters put both lines of thinking into serious doubt. Empirical indications suggest that algorithmic VOD recommender systems remain largely complementary, blunt instruments of suggestion. The data collected in my surveys and interviews are no outliers; my conclusions confirm, complexify, and update prior studies conducted in other countries. Hardly confined to the United States or United Kingdom, the phenomena illustrated here are anticipated by international surveys that cast doubt on the importance of Netflix's and competitors' recommendations to users' consumption outcomes. A 2014 large-scale study of film audiences in ten European Union countries demonstrated, for example, how tips from friends and family remain the most relevant source for viewing decisions, regardless of whether film is to be consumed at the cinema, on television, or via VOD. For cinema viewings, word of mouth remained "very important" or "important" to 73 percent of surveyed Europeans, and as such the most important source.[4]

When asked specifically about films watched on VOD services, the EU study participants ranked word of mouth as far more important than recommender systems (and their practical expressions and proxies, such as "prominence on platform," "tailored recommendation," or "editor's choice"). Of the respondents, 66 percent rated prior knowledge (film I have already heard about) as very important or important, compared with genre search (65%), word of mouth from friends and family (64%), film trailer (on platform) (59%), joint decision while consuming VOD in a group (56%), tailored recommendation (based on prior viewing behavior) (54%), freshness (titled added most recently to catalog) (52%), popularity (most watched titles) (51%), user ratings, votes, and comments (47%), recommendation of friends via social networks (45%), exclusivity (title available nowhere else) (41%), prominence on interface (39%), and editor's choice (titles recommended by the service) (34%).[5] In sum—in a year in which Netflix captured 66 percent of the world SVOD market and in a duopoly together with Amazon captured nearly 94 percent market share in some countries included in the study, including the United Kingdom—VOD recommender systems ranked among viewers' *least* influential forms of cultural mediation.[6]

Anticipating my own results, the 2014 EU survey indicates yet again that traditional forms of information, especially word of mouth, are significantly more decisive in forming opinions about what to watch, rather than, as some commentators assert or imply, algorithms.

Confirming decades of research from a whole host of disciplines, fields, and geographical regions, this book has shown that in general—even in the online digital age of aggregation and algorithms—most people still prefer alternatives to VOD recommender systems when choosing films and series. By all measures, traditional word of mouth remains the most trusted, most "personalized," and most prevalent form of cultural recommendation. Despite the remarkable growth in streaming, the vast majority of film and series consumers do not use VOD recommender systems as their sole or even primary source for viewing suggestions. The weight of evidence from both my empirical project and others suggests that these tools may be less influential than their proponents *and* critics believe.

Forget for a moment the mouthfuls of algorithm types, the Markov chains and restricted Boltzmann machines, and compare a Netflix recommendation—from a user's perspective—with even a relatively simple word-of-mouth encounter about media products, which, according to research, we are having 2.8 times per day on average.[7] Let us imagine such a situation. Friend A gives Friend B a tip about a series she thinks Friend B will enjoy, *Deadwood* (2004–2006). Friend B is intrigued and, not having heard of the title before, asks what the story is about. It's set in the 1870s US-American frontier, Friend A begins, and stars Brian Cox, a favorite character actor of Friend B. Friend B interrupts and remarks that he is not much of a fan of westerns or period dramas. Friend A persists, countering that *Deadwood* is unlike any western she has ever seen. The writing reminds her of *The Sopranos* (1999–2007) and the storyline and ensemble cast recall *The Wire* (2002–2008). Since these references appeal to Friend B and because Friend A has provided good tips in films and series in the past, Friend B overcomes his initial skepticism and endeavors to seek out *Deadwood* at the next opportunity.

This word-of-mouth scenario, which most would accept as ordinary and quotidian, demonstrates the severe constraints on Netflix's recommendation engine: the latter's suggestion of *Deadwood*, if received without any other information, would most likely die at the point that our Friend B sees the description of 1870s US-American frontier, if not already upon scanning the corresponding cover art of men in cowboy hats. (The examples from chapter 5 demonstrate as much.) Netflix's recommendation engine, for all of its personalization inputs, compares poorly with the dynamic, dialogic nature of most word-of-mouth situations, which can provide persuasive personalized contexts, as here with the references to *The Sopranos* and *The Wire*. Unless the service develops such a high credibility with the individual user whose experiences of Netflix's suggestions are reliably suitable, Friend B will not be able to overcome his skepticism, his "adoption threshold," derived from his initial superficial perception of the title.[8]

Indeed, beyond such thought experiments, studies show that although algorithmic recommender systems may be in general and on average more precise than humans in predicting the preferences of others, certain individuals outperform the machines easily and substantially.[9] Because of their heavy use and sophisticated film knowledge, these human "superpredictors" understand the quality and attributes of particular exemplars. Recommendation engines, of course, for all of their technological complexity cannot know that any film or series is in fact a dud despite a few initial ratings that might suggest otherwise. Let us call these prescient predictors WOM Providers, your clever culture-vulture friend, or, indeed, trusted critics.

Word of mouth is an intensely personalized form of recommendation. Participant 25, who cannot bear spoilers, relies heavily on film tips from his father and close friends because of this fact: "I manage to avoid any spoilers by making sure that friends know. My dad, for example, knows that I hate spoilers, so he will make sure that he doesn't tell me anything. He'll just tell me the surface of the story or whatever." There is no doubt that people may interact with computers as "social actors,"[10] maintaining parasocial relationships with Netflix and its recommendation engine. Chapter 5 illustrated how some clearly do so more than others. And yet human interactions, the interpersonal intimacy of exchanging tips with friends, family, colleagues, and acquaintances, while potentially more time-consuming, remain immensely gratifying forms of cultural mediation. Word of mouth has social benefits to both the provider and the recipient: humans enjoy sharing stories and reviews, and recommending an interesting film or series highlights the provider's extraordinary knowledge and experience in an act of social confirmation and approval.[11] After all, our consumption of films and series is often not an end to itself, but rather a means of participating in, and announcing our position in, social life: "Keeping Up."

My findings do not dispute the complexity of Netflix's or Amazon's code and inputs. It is undoubtedly true, as one Netflix engineer boasts, that "there's a whole lot of Ph.D.-level math and statistics involved" in the company's online data collection and algorithmic processing.[12] And yet BuzzFeed's "Netflix in Real Life" video and other pop-culture expressions and memes spoof the system as decidedly fallible and wooden in practice. Many of my interviewees—including those who use such tools frequently—rated the quality of VOD recommendation outcomes as very poor. A friend providing such low-quality tips would no doubt lose credibility quickly. Both the cheerleaders and the fearmongers of the internet and big-data "revolution" falsely equate technical complexity with power and credibility among users.

With further developments in data engineering, Cinematch and other VOD recommender systems will no doubt improve in years to come. Inversely to the case of Participant 13, whose trust in Netflix has eroded because of poor tips, a future technological breakthrough may bolster VOD suggestions' credibility. But there are also indications that this may not be the case: Facebook, Google, and other algorithmic search-and-recommendation engines, after a starry-eyed honeymoon period, have suffered declines in reputation and trust, despite technological enhancements.[13] The novelty has evaporated and, for some users, an ambivalent "don't-care" cynicism is setting in; among millennials and other digi-natives, a blasé attitude toward personalization in cultural pursuits abounds.[14] Although both the EU survey (fieldwork undertaken in early 2013 and published in 2014) and my own 2018–2019 empirical audience study found recommender systems to suffer from relatively low credibility, the downward trajectory between those results over five years—a period in which Netflix and a few other streaming providers became mainstream and suspicions about algorithms proliferated in the news media—suggests that credibility may be *declining*, not improving, the more that users engage with these devices.

In any event, I struggle to imagine a future where such recommender systems have simply replaced word of mouth, mass-market ads, trailers, text-based reviews, and other traditional means to make cultural consumption choices. It would seem that even the smartest data engineers, using the most ingenious algorithms and the largest imaginable volumes of data would fail to always and in every instance perfectly catch the mood and whim of individual members of a notoriously capricious and self-centered species. Some observers worry that the Netflix recommender system's "new meanings and practices can insinuate themselves into long-established routines, transforming the latter in ways that may be just reaching popular awareness."[15] However, this book has outlined a host of reasons to doubt the extent to which Cinematch, or any other competitor for that matter, will substantially redefine culture and taste.

We must be careful not to caricaturize Netflix's recommender system and other algorithm-based cultural mediators as all-knowing, all-powerful systems that can singlehandedly alter tastes and profoundly change the habits of defenseless or ignorant consumers. This sort of critique inadvertently aggrandizes such systems at the same time that it unwittingly participates in a "media harm" discourse: Horkheimer and Adorno for the Silicon Valley set. We need to relinquish purely techno-deterministic arguments: that because of the recommender system's technological design, it will (or even has the power to) a priori transform users into Adam Sandler–gorging sloths, lowest-common-denominator gluttons. Indeed, we cannot assume that any given technology has any certain effect on any one user, on all users, or indeed on culture at large. Although a structure, form, or affordance may enable, disable, incentivize, or disincentive certain effects, these outcomes are not foreordained, nor will they be the same for everyone.

Ramon Lobato has noted how Netflix, although surely a global business and social phenomenon, serves diverse purposes and clienteles across its many territories of use. The company "has had quite different effects in different national contexts," competing with different rivals and attracting different types and quantities of subscribers.[16] In equal measure, one must conclude that the *uses* of Netflix and VOD recommender systems are hardly uniform, functioning within subscribers' unique media repertoires and in various constellations with other information sources and consumption channels. As an intervention into research on digital-age media audiences, this book challenges prevailing sweeping diagnoses of either a "participatory culture"—the belief that new digital technologies lead to more active, empowered audiences—or an equally immutable "algorithmic culture," by which user agency evaporates under the onslaught of mathematical formulas and capitalist-informed personalization. Neither narrative adequately accounts for media use in the age of Netflix.

There is no "typical" Netflix consumer. Media users are diverse and different. Of course, there are some who are misinformed or ignorant, blithely unaware of the technologies and agendas shaping their viewing habits. Surely, we must interrogate how VOD services veil or rationalize these ulterior motivations in their interface design, promotional rhetoric, and recommendation styles. And yet we must recognize that such ignorance is neither total nor ubiquitous: users' sophistication in understanding VOD and its dynamics of suggestion varies widely. When we talk about obtrusiveness, transparency, and knowledge vis-à-vis media technologies, we must question generalizing assumptions and take into account users' capabilities and concerns: what is unobtrusive for some may be obtrusive for others. Indeed, the evidence presented in this book indicates that many if not most users see through the opacity to some degree, maintaining a functional understanding of these systems. Whether or not this knowledge changes their behaviors is another matter altogether, as we have seen with the "don't-care" faction.

To wit, we must differentiate user intentions: how one encounters media and for what reasons one approaches them. The long history of empirical research, as presented in previous chapters, shows that media choices are complex. One's position on the "openness" scale, for example, affects how inclined one is to seek out and accept novel, diverse, or serendipitous items. Levels of "conscientiousness" and the extent to which one is a "maximizer" or "satisficer" influence how much (or little) time one will be willing to spend scrolling through rows. Agreeable users may be more likely to trust different forms of recommendation. Personality type, level of media use, and the perceived stakes of that use are among many lesser-studied contingent factors that research has shown to be just as if not more predictive of entertainment preferences than traditional demographic indicators.[17]

Finally, Netflix and its critics assume not only that Cinematch suggestions are sophisticated, well-tailored to individual tastes, credible, and thus being followed. Media scholars' "posthuman" arguments rely on another supposition: that everybody is doing it. Such theories might be sustained if *all* consumers *only* used algorithmic VOD recommender systems and if all algorithmic VOD recommender systems were designed to circumscribe tastes in a particular way. Indeed, those who predict Netflix's or Amazon's wholesale hijacking of audiovisual culture must acknowledge that SVOD services with algorithmic recommender systems may be the fastest-growing means of media consumption, but writ large, they still command only a minority of consumption, for most people only a small proportion of their overall media diets. Despite the many obituaries, linear television still represents by far the most popular locus of audiovisual consumption; most people partake of a wide range of content delivery options.[18] There remains—and by all prudent predictions there will remain for a considerable period—an overlap of delivery systems. As a number of scholars have pointed out, despite social herding and the widespread promotion of novelty in publicity and the press, the adoption of new media is never simultaneous or seamless: emergent, dominant, and residual media coexist and commingle.[19] Media choices intersect. Despite widespread anxieties in earlier eras, television did not kill movie theaters any more than video rentals fully substituted for television.[20] The enduring existence of cinemas shows how older delivery systems maintain intimate relationships with newer varieties, often declining in use but just as often transforming when alternative media optimize and take over certain functions. The mere existence of algorithmic recommender systems does not constitute a wholesale dumbing-down or quantification, but rather a partial repurposing and diversification, of taste and mediation cultures.

Thus far, Netflix and other VOD services may partly displace, but do not entirely replace, other means of media delivery. Very few if any consumers *only* watch films and series via SVOD and even fewer watch *only one* SVOD platform: in the United Kingdom, for example, people already subscribe to Netflix *and* Amazon much more frequently than they do to either alone.[21] Indeed, those who engage with VOD most—the supposed victims of algorithmic filter bubbles—are precisely those consumers most likely to be omnivorous in regard to delivery methods and portals.[22] In both the United Kingdom and the United States, those most likely to use streaming, and to stream frequently, are also the same people who are most likely to go to the cinema (frequently), watch broadcast television (often), and so on.[23] The vast majority of consumers are spending the vast majority of their film and series time with a delivery mechanism, linear television, into which algorithmic recommender systems do not directly intervene. Despite the heavy attention afforded to the new "quality" content on VOD and premium cable, whether *House of Cards* (2013–2018) or *Game of Thrones* (2011–2019), audience figures for these programs pale in comparison to broadcast offerings such as mid-season episodes of *The Great British Sewing Bee* (2013–2016) or *Big Bang Theory* (2007–2019). Amid the heady talk about filter bubbles, fragmentation, and on-demand culture, aggregated television viewership actually remains on the rise. Linear, broadcast television retains a significant, indeed dominant following.[24]

Even these conclusions, as careful, tentative, and gray as they are, assume that SVOD viewership will steadily increase as a proportion of viewers' stagnating media time and that Netflix and Amazon will continue to gobble up market share. And yet we know that consumers increasingly subscribe to multiple SVOD services, including some platforms much more invested in content diversity than Netflix. In this way, these consumers are self-diversifying, compensating for Netflix's dwindling catalog by seeking out exposure to alternative content offerings. We need to see these services as complementary; consumers cancel, shift, or double up as individual providers fail to offer the one stop shop that early prognosticators foresaw for SVOD.[25] "Real pluralism," as one scholar states, "is only obtained when diversified content is actually consumed."[26] At least some media consumers, as chapter 5 has illustrated, perform this task intuitively: they exert their free will and compensate for the flaws and lacunae of Netflix and other media portals. A user-centric perspective, such as practiced in this book, can enrich studies of technology and form and generate more nuanced assessments of cultural effects.

THE PERSISTENT THIRST FOR GATEKEEPERS

In the end, the folk beliefs and scholarly debates surrounding recommender systems are compelling less for the ways in which their respective logics can be put under scrutiny or "disproved" than for what they tell us about deeper social needs and desires. Mindful of Mosco's analysis of the broader rhetoric surrounding the introduction of new media, we need to demystify the fantasies and fears underlying these specific areas of film and television culture. We must also remain cognizant of how these ideas and assumptions about taste and technology might configure in hierarchies of power. Both real users' and commentators' *functional* articulations about cultural mediation revolve around aspirations for choice, on the one hand, and a cohesive public sphere with benign yet strong gatekeeping, on the other. The latter has been one of the most persistent and longstanding cultural yearnings since at least the Victorian era.[27]

Yes, internet-inspired kill-the-gatekeeper rhetoric and public antipathy toward experts may be increasingly prevalent and socially acceptable. Democratized (and often virulent) expressions of taste proliferate in online forums. And yet my research suggests that the need for cultural mediation— whether in the form of Netflix and Amazon algorithms, MUBI and BFI Player curation, Rotten Tomato quotients, or traditional word of mouth— has only increased in direct relation to the digital-age content explosion. Real users, chapter 5 has demonstrated, feel overwhelmed by the proliferation of offerings. Just as ever in the history of cultural recommendation, some are willing to invest time and money engaging with complex recommendations and thorough explanations and justifications. Many more, however, remain satisfied (or quietly dissatisfied) with quick tips.

Despite widespread promises of disintermediation—the hope (or fear) that content producers would "cut out the middleman" and offer their products directly to consumers—gatekeeping, in the sense of limiting and prioritizing content, endures. Indeed, reliable information regarding the proliferating range of media texts is more valuable than ever. Intermediation has always been indispensable to reduce consumers' (opportunity) costs of seeking out and evaluating experience goods such as films and series. Despite and because of the increased availability of content, despite and because of little blue links on dodgy BitTorrent sites in the netherworlds of the internet, guides remain essential. In a media world where barriers to production have become exceptionally low, where governments use cultural industries—via tax breaks and direct subsidies—to stimulate economic activity and promote tourism, and where new VOD services aim to thrive with talked-about content, films and series are multiplying annually. Faced with a proliferating industry of prototypes, unique objects requiring advance information, humans must employ heuristics and repertoires; they need filters, labels, and "infomediaries."[28] Now perhaps more than ever, cultural recommendation remains necessary for a successful and satisfying participation in media life. To the extent that disintermediation has occurred, it is rather a matter of *different* middlemen of altered type and location: Amazon and iTunes represent the interfaces through which consumers purchase or rent videos, rather than Blockbuster and Walmart, for example. Companies like Google, Facebook, MUBI, Rotten Tomatoes, and Netflix have become key content aggregators, curators, and even producers.

Rather than disappearing, gatekeepers now need their own gatekeepers: recommender systems now invite their own recommender systems. For evidence one must look no further than Rotten Tomatoes, the early days of Jinni or Watchly, Flickmetrix, Film Affinity, or the *New York Times*' Watching. These services represent meta-recommender systems. Watching asks users to select a few attributes to describe the content they seek (Inspirational, Dark, Binge-Worthy, Joke-Heavy, Strong Female Lead, and so on) and, if necessary, to further refine the suggestion by selecting a genre (e.g., Comedy, Drama, Animated, Reality, Thriller) and a subgenre. The generated list provides films or series that fit the viewer's mood, including thumbnail images, a pithy plot and genre overview, a descriptive rationale for and against watching the selected content, the Metacritic score, and, perhaps most importantly, the VOD services that currently stream the content. Watching is suggestive of the present and future of audiovisual-product recommender systems for a number of reasons. It specifically pitches itself as a recommender for the moment and "present self," rather than necessarily great works for the aspirational "future self." (The first stage of the recommendation process makes this explicit in its question: "What are you in the mood to watch?") It also attempts to address users' oft-stated objection that algorithms cannot "know how I'm feeling that evening." Computer programmers categorize such tools as conversational, knowledge-based, or context-aware recommender systems.[29] Far from eliminating gatekeepers, Watching adds another layer, articulating its value as a one stop shop, freeing the consumer from having to check two, three, four, five, or more different services.

In general, and beyond VOD services themselves, rather than any disintermediation there has been a wholescale *reintermediation* of culture: the aggregation, filtering, and curation of content under a proliferating amount of unequally used providers, with recommender systems complementing traditional forms of suggestion and evaluation. Google, iTunes, Zagat, a favorite blogger or a YouTube influencer, Yelp, Reddit, BuzzFeed, TripAdvisor: these are the new, and not so new, middlemen that now compete with traditional critics' reviews, word of mouth, and all manner of advertising. Even (and perhaps especially) in the digital age, filters color our tastes: we view the cultural world through intermediaries and yet we often fail to recognize them. Pierre Bourdieu once argued—writing in the 1970s and thinking of human critics—that the role of cultural recommendation in structuring taste is such an invisible and quotidian part of life that people tend to believe their choices and preferences are wholly their own.[30] This observation obtains also today in the face of interfaces and rhetoric that only serve to confirm this myth.

DEATH OF THE CRITIC?

By way of conclusion to my study's core issues, let us return to grapple with arguments that algorithms attempt to displace learned expertise (e.g., critics) with numerical, and thus dumbed-down, forms of cultural recommendation. Can Netflix be considered as a form of criticism, in the long and varied spectrum from catalog synopses and capsule reviews to academic essays and books? It is clear that, from the perspective of the user, VOD recommender systems exist on an axis of information and suggestion sources, including ads, Tweets, trailers, user posts, word of mouth, and considered reviews.

To be sure, the Netflix recommender system competes with critics' reviews in the sense that both function to recommend, inform, and provide context. Some consumers, incentivized by the system's seamless incorporation into the Netflix interface, may be less likely to read a review, phone a friend, or consult information on IMDb, Rotten Tomatoes, or elsewhere. As we have seen, consumers engaging with media for casual purposes will often attempt to bypass cumbersome or time-intensive means of investigating film and series choices. Looking up a review in a stack of old newspapers or a table-side copy of *Leonard Maltin's Movie Guide* were relatively inefficient processes for snap consumption decisions. By outsourcing the research into what films or series might appeal to them, viewers save considerable time and effort.

Nevertheless, audience studies indicate that such facts hardly entail the end of the learned critic. The 2014 EU survey of VOD users suggests that learning about a film beforehand (e.g., via critics' reviews) is a significantly greater inducement to watch content than a tailored recommendation based on prior viewing history (66% to 54%; "very important," 22% to 15%). My own study indicates that consumers use a wide variety of means to make decisions about consuming films and series; the survey results indicated that US and UK adults report using reviews significantly more often than recommender systems—and that they trust the former significantly more than the latter to provide a satisfying film or series tip. The qualitative analysis of interviews demonstrated how even the decision to click on a film or series title in a recommended row will often incorporate several means of suggestion in an iterative process that may span minutes, hours, weeks, or months: for example, the memories of posters, a friend's tip, or a review scanned months prior. The decision to watch comes only after exceeding any given user's content-specific quantitative threshold of recommendations and a qualitative threshold of recommendation-source credibility.

Yes, human-generated reviews remain a secondary reference point for most viewers. As confirmed in the fine-grained results of my survey, consulting criticism is a part of a specific media repertoire of a particular subgroup: traditionally middle-class, heavy media users, and at any rate those seeking a more intense engagement with particular content. In fact, film criticism has never enjoyed the sort of influence and widespread use that some pundits imagined. Decades of audience studies from the fields of film and television, communications, economics, marketing, and social psychology have consistently demonstrated that critics overestimate their own importance in terms of short-term tastemaking and box-office influence and underestimate their resilience in long-term canon building and other social functions: testing opinions; experiencing an imagined cultural dialogue; affirming, guiding, or legitimating one's own taste.[31]

Commentators are perhaps correct to claim, moreover, that the Netflix recommendation engine is in step with a larger shift in discourse about taste: from what is objectively and universally good, an Arnoldian method of core-text selection (still practiced to a great extent by universities and by museums like the Louvre) to one based on a selection of personalized preferences that may or may not be applicable to others. No longer are we at the mercy of critics and scholars and their visions of the best that has been thought, said, and recorded: an empire of "likes" and constellations of little stars now inflect the consumption of moving images. Netflix does not try to tell you what are the best works; it tells you what best works for you.

The decay of great-works Arnoldianism is a long time coming, however, and hardly a unique characteristic of Netflix and other algorithmic recommenders. That paradigm depended on a small, homogenous cultural elite and a correspondingly circumscribed audience for discerning cultural products. In that age criticism strongly resembled word of mouth, so tightly knit were the closed-shop networks of boutique periodicals, and the inbred coteries of their messengers and recipients. Today, recommendation engines are contributing to a reconfiguration of consumption-decision protocols. But rather than a wholesale redefinition of the public sphere, this represents a partial individualization of culture, an incremental validation of unique tastes and anti-hierarchical cultural relativism: phenomena in motion for at least the last fifty years.

Ultimately, Netflix is not in the business of *understanding* the larger coordinates of personal taste, quantitatively or otherwise: *why*, for instance, I enjoy thrillers set in New York, but my wife likes 1980s horror and coming-of-age films. Rather, Netflix and other competitor recommenders remain much more invested in microlevel *whats* of taste (*House of Cards* viewers tend to like *It's Always Sunny in Philadelphia* [2005–present]) and on aggregate-level *hows* (which it claims to use in commissioning original content). Netflix is no analyst trying to explain what pleasure or utility we derive from a slasher film or a romantic comedy. After all, thousands of academic articles and books exist precisely for this purpose.

A Netflix recommendation is certainly reductive when compared to insightful critics' winding, reflexive, informed, free-wheeling assessments. Then again, this latter sort of criticism constitutes the tiniest sliver of film and culture evaluation. Indeed, Hallinan and Striphas, in their efforts to diagnose algorithmic recommender systems as symptoms of a cultural decline, elide an essential fact: the description "written argumentation" flatters the vast majority of human-generated film appraisals. Well before the advent of Netflix and Rotten Tomatoes, critics reflected on thumb- and star-evaluations as symbols of a professional dumbing down.[32] Arguments that Netflix's behavior-informed suggestions constitute, *partout*, an evacuation of humanistic culture presume that no one will ever read another review (or look at billboards or take advice from friends) again. This is clearly not the case. In general, commentators' objections resemble the "death-of-the-critic" thesis: a depreciation of cultural capital in the face of new media.[33] Their work recalls past iterations of nostalgia for fading media routines and repertoires. Common to these projects is a mourning for the perceived loss of gatekeepers and human qualitative descriptions, elegies for public critics and prominent intellectuals, and pronouncements of a humanities in crisis.

Algorithmic recommender systems may well appropriate some aspects and forms of criticism, but they do not wholly substitute for critics—either in a cultural or an economic sense. Hitherto, they are no match for finely expressed, perceptive, human-generated criticism. With minimalist explanations like "86 percent" and "Because You Watched," Netflix recommendations favorably compete only with broadcasted thumbs-up, thumbs-down puff pieces and summary-heavy capsule reviews. In various iterations since the turn of the century, Cinematch has only provided the bare minimum to answer the basic reviewing questions: What kind of thing is this? What is it about? To what is it similar?[34] Beyond the single referent provided in the "Because You Watched" row, Netflix essentially capitulates on criticism's fundamental requirement of providing "evaluation grounded in reasons." If the metaphor is to hold at all, then let us consider the algorithmic recommender system as an automated option in bottom-barrel, quick-fix cultural decision-making situations. It operates at a low functional level; as we have seen, its use value remains circumscribed for most. In this sense, *pace* Striphas et alia, Netflix remediates but does not replace considered criticism. If anything, recommender systems supersede the paper search of Leonard Maltin's tome of blithe capsule reviews or the gloss of the Thursday newspaper listings; obviate the laziest phone call or text suggestion from a friend; supplant the "staff recommends" shelf at the local video shop.

In most cases, however, recommender systems replace nothing at all. The empirical evidence intimates that many who use Netflix suggestions in a one-stop-shop fashion as their *only* information source for a consumption decision would have probably consulted no further details, beyond the onscreen guide, if the program had appeared during channel surfing. In prior eras, many may not have even watched a film or series in such situations. Despite its technological sophistication and claims of personalization, Netflix's recommender system represents a low-level, low-risk form of endorsement; suggestions are embedded within a content delivery system of immediate gratification and an all-you-can-watch subscription economics with almost zero opportunity costs for sampling (bad) content. Real users report that their *sole* reliance on Netflix's recommender system usually corresponds to a low-stakes engagement: for example, passively viewing while texting or background noise while cooking. In some ways, this should not surprise us. After all, computer scientists foresee recommender systems as "tools for helping people to make better choices—not large, complex choices, such as where to build an airport, but small- to medium-sized choices that people make every day: what products to buy, which documents to read, which people to contact."[35] In contrast, those who seek out films or series for intense emotional or intellectual engagement, in situations whereby audiovisual content means more than mere background noise, remain as likely to read reviews or seek further information after consuming the content. Theories of a steadily encroaching algorithmic culture eclipsing human criticism and cultural evaluation simplify a much more subtle phenomenon.

Even among VOD services there is a significant (albeit minority) subculture that eschews the big-data approach, presenting content together with criticism and criticism-like recommendation forms. The backlash against algorithmic recommendation, epitomized in providers such as MUBI, indicates that while certain audiences will tolerate or even welcome wholly data-driven suggestions, others (heavy users and those interested in elaborated context and varied forms of content) will probably always find them inadequate. Curation-style services such as BFI Player, Kanopy, or FilmDoo reach (those relatively few) users who want much more than a quick tip. They promote themselves on their boutique selections, bespoke service, thick descriptions, in-house critical organs, social community, high-quality content, and pretentions to distinction—not simply different in terms of genre variety, but moreover an unknown or exotic culture, a new vision, perspective, or means of expression. In light of such platforms, and the fact that algorithm-led recommender systems are neither used universally nor regarded as having high levels of credibility, it would seem premature to make all-encompassing pronouncements about the end of expertise.

It is far from a foregone conclusion that algorithms will "solve" film and series recommendation—the "nobody knows" problem amplified in an age of excess—in the way that the compass resolved implacable risks of ship navigation. Companies that furnish the best offer—quality as much as quantity, service as much as content, presentation as much as selection, and possibly community and context as much as personalization—may thrive. But it is just as likely that we may arrive at a certain dominant configuration because of inertia, rather than on the basis of market logics or the satisfaction of consumer demand. Will the dominant means of VOD recommendation be partly, mostly, or exclusively algorithmic? Will it be premised on the promise of curation and human expertise? Gut instincts, reasoned arguments, or the putatively democratic wisdom of crowds? Will the ultimate sources of these recommendations be industry executives, skilled programmers, seasoned critics, or regular Joes? Will the suggestions be shared or broadcasted, personalized or private? Coupled with exclusive programming or tie-ins to hardware, cinema tickets, or meetup social clubs? It is already evident that there is no single answer for the entire market, even if it remains equally clear that certain forms, above all algorithmic recommenders and original programming, are thus far proving to be more compelling business propositions than others. The history of new media demonstrates that what in hindsight may seem like inevitable conclusions were in fact cases of "settling" on a certain outcome: the more any given company deploys one technology or form over another invites imitation, which in turn creates industry myths about consumer preferences and catalyzes economic incentives to coalesce around the "winning" format, model, or technology.[36] Despite ex post facto journalistic framings, narratives of heroic and inevitable "innovation" mask social herding; pragmatic, defensive, and reactive corporate maneuvers; and a more or less arbitrary consensus.

A quarter century after the launch of Netflix, in a digital age that is no longer novel but has arrived at a consolidation stage, there is no question about the existence of, and social need for, sources and tools to guide film and series choice. Although the speed and location of access have intensified and dispersed, respectively, the availability of various possible influences is not new. The historical lessons of new media also apply here: recommender systems remediate predecessor forms by "presenting themselves as refashioned and improved versions."[37] Rather than seeing such tools as end states, we need to consider them as evolving forms with pasts and futures that seek to address long-standing human desires and commercial mandates. This book has demonstrated how the current battle over cultural recommendation is not merely a competition of systems or even technologies, but moreover one of marketing, of rhetoric, of credibility, of user agency and folk beliefs: of how the cultural world should be structured and who should determine its designs, functions, and underlying content.

ON METHODS AND NUMBER TROUBLES

The public discussion of recommender systems must be undertaken with caution, care, and calm. Above all, it must not be reduced, as it hitherto often has been, to a tribalistic feud between the quantitative and the qualitative, a long-running and shortsighted guerilla campaign of the "two cultures" paradigm demanding impenetrable firewalls between science and the humanities.[38] Thus far, the debate resonates with a larger, long-standing antipathy among humanist scholars toward quantitative measurements and computational culture. According to danah boyd and Kate Crawford, "Big Data risks reinscribing established divisions in the long running debates about scientific method and the legitimacy of social science and humanistic inquiry"—that is, the "mistaken belief that qualitative researchers are in the business of interpreting and quantitative researchers are in the business of producing facts."[39] Wanting to unite rather than divide, I would avoid notions that create a dichotomy between supposedly quantitative and qualitative notions of taste, that imply diametrical oppositions and ontological incompatibility, or that ring-fence research areas, terms, or ideas (taste, identity, emotion, and so on) as out of bounds for one method or another. This book's intentionally mixed-method approach eschews illusions of qualitative-quantitative purity, unproductive disciplinary feuds, and boundary policing.

Each side has at times perpetuated caricatures of the other.[40] But those of us trained primarily in qualitative methods must also critically examine our own biases before casting stones with charges of unreflective, reductive positivism. Rather than fostering debate, there is a danger that such epithets foreclose it by insisting that numbers could not, in any circumstance or fashion, play a role in assessing matters of taste. (After all, Pierre Bourdieu's *Distinction*, on whose authority some of these writers rely, is itself an empirical study based on quantitative survey analysis: in a literal sense, a mathematics of taste.) Whether in thick descriptions or in quantitative approximations (or vice versa or both), scholarship should aim to make the world intelligible. Numbers and algorithms deserve the same cultural scrutiny as words and discourse; and yet neither, to paraphrase Steven Pinker, should be the a priori enemies of humanities.[41]

A mutually inclusive, hybrid approach seems beneficial as a methodological framework on its own terms. Such a procedure also corresponds to how recommender systems in fact straddle supposedly tidy boundaries. This book has shown that behind the swagger and PR drivel, the lines between systems' qualitative, human, and editorial characteristics, on the one hand, and the quantitative, computational, and data-driven features, on the other, are blurrier than they may first appear. On the one hand, MUBI's "curated" recommendations may not be algorithm-driven, but they are certainly data-informed. On the other, there is significant backroom backtracking from the ideological purism of machine learning over human expertise. Amazon has purchased GoodReads, a user forum for decidedly qualitative reviews, as a horizontally integrated supplement and complement to its algorithmic recommender systems for books. Apple has built a human newsroom of thirty former journalists who spend their days curating a newsfeed from one hundred to two hundred pitches they receive from publishers, deciding which stories are important enough to appear on the top of lists.[42]

Even Netflix, for both cultural and economic reasons, has quietly retreated from its algorithm-above-all approach. As the company has plunged into original productions—and has begun to function more like a major studio or television network than an upstart tech company—the Hollywood-Silicon Valley balance has skewed more and more in the favor of the former. Although engineers advise that a certain image or star may trend better in A/B tests of the recommender system, executives curtail and increasingly overrule the data-focused approach, learning—in the words of a *Wall Street Journal* exposé—to "temper [the company's traditional] love of data models and cater to the wishes of A-listers and image-conscious talent, even when they might be at odds with the 'algorithm.'" Netflix has ceded to the wishes of powerful producers and actors. In order to grease professional relationships, honor contracts, or defuse delicate PR issues, it renewed *GLOW* (2017–2019) and *Lady Dynamite* (2016–2017) and changed the personalized trailers, images, and other marketing efforts for Adam Sandler and Jane Fonda vehicles, even though the recommender system's data suggested that other tactics would attract more eyeballs and better serve its customers.[43]

These developments follow on from the company's epiphany, examined earlier in this book, that its "algorithmic" recommender system required hiring *humans* to tag factual information and evaluate emotional valences.

These less publicized business moves confirm a fact often lost behind Netflix's and competitors' bombastic Wizard-of-Oz rhetoric: the technology behind recommender systems is not incredibly effective. Indeed, even computer scientists admit that, despite tech-company slogans, AI is still in its infancy. According to one, the "challenge of creating humanlike intelligence in machines remains greatly underestimated. Today's A.I. systems sorely lack the essence of human intelligence: *understanding* the situations we experience, being able to grasp their meaning."[44] These systems, as real users complain, seem incapable of understanding how one is feeling on any given evening. Many algorithmic programs—including high-stakes applications deciding the fate of parolees—have the same rate of success and failure as *untrained* humans.[45] Despite the incredible capacity to aggregate taste choices quickly and at large scale, algorithmic computational processing lacks basic common sense, let alone the ability to reason. "Stop thinking about robots taking over," computer scientist Barbara Grosz admonishes: "We have more to fear from dumb systems that people *think* are smart than from intelligent systems that know their limits."[46]

Pointing out the flaws of AI, algorithmic computation, and VOD recommender systems—and thus the limitations of many prior humanistic critiques of these instruments—does not constitute a call to obviate media ethics or discontinue critical reckonings with big tech in the cultural sphere. Rather, this book recommends executing this critique soberly and with an informed, rather than inferred, understanding of its role in users' lives.

ROBOT CRITICS AND THE AUTOMATION OF THE CULTURAL PROFESSIONS

The evidence laid out in this book suggests that humanists may have less to fear from VOD recommender systems than hitherto supposed, because such tools have clear historical precedents, and are not nearly as novel, effective, or widely used as claimed. Such conclusions may intimate an altogether sanguine outlook, a "don't-worry-be-happy" harmony between more or less human-led and more or less algorithmic cultural recommendation. This interpretation, however, would neglect the nuance of the presented facts.

The thesis that recommender systems are not replacing humanistic culture wholesale can obtain in a world where these developments may nonetheless exert economic consequences—in the form of inequities of wealth, job losses, and casualization or reduced pay—for the human practitioners of cultural intermediation. The findings I submit here do not preclude any and all effects on labor relations, on individual careers and livelihoods. Indeed, commentators' quantitative-qualitative quandaries symptomatically gesture to more existential anxieties about how incipient forms of automation in the cultural sphere may herald the loss of authority, status, purpose, and economic sustainability of the *practitioners* of humanistic intermediation.

Of course, we need to face some uncomfortable facts and less rosy assessments. Dark predictions of a "useless class" of workers and regular reports about automation reducing needs for human labor abound, with Amazon functioning as a prime exhibit in the deployment of robots and drones.[47] According to PwC, ten million workers in the United Kingdom are at risk of being made redundant in next fifteen years, including 2.25 million in the retail and distribution sectors.[48] In turn, Oxford academics Carl Benedikt Frey and Michael A. Osborne calculate that 47 percent of jobs in the United States could be automated by 2023.[49]

Automation has replaced humans in many industries. We tend to see this as triumph-of-science progress and the end of backbreaking drudgery, not to mention the anomie of repetition; up to now, the salaried classes have largely benefitted from these innovations. But wearing a white collar to work or receiving a fixed monthly wage does not inoculate one against redundancy. Indeed, reports suggest that the main locus for automation has moved beyond low-skilled manual labor and "routine manufacturing tasks," into middle-class domains.[50] Any professions or duties not requiring complex social interactions, dexterity, or empathy, experts claim, are prone to replacement.[51] Computational processes can now replace traditional human "knowledge work," one McKinsey review admits, by automating and personalizing alerts to consumers about what they want to buy.[52] "Occupations that require subtle judgment," according to Frey and Osborne, "are also increasingly susceptible to computerisation." In these professions, which include judges—who must interpret facts against preexisting guidelines and among whom, studies show, human biases run rampant—or social workers who must detect child abuse, "the unbiased decision making of an algorithm represents a comparative advantage over human operators."[53]

Much literature not only forecasts the automation of labor but also refers to these developments as foregone conclusions, in the present or even past tense: Christopher Steiner's *Automate This: How Algorithms Took Over Our Markets, Our Jobs, and the World* and Alex Rosenblat's *Uberland: How Algorithms Are Rewriting the Rules of Work*.[54] According to Steiner's alarmist rhetoric, algorithms "normally behave as they're designed. . . . But left unsupervised, algorithms can and will do strange things. As we put more and more of our world under the control of algorithms, we can lose track of who—or what—is pulling the strings. . . . algorithms are taking over everything."[55]

These discussions are clearly causing anxiety among the populace overall. Pew Research Center data reveal that 72 percent of Americans worry about robots, computers, and automated processes replacing humans.[56] News items on angry mobs wielding rocks and knives in guerilla attacks on self-driving cars suggest popular unrest with the perceived dangers of machine learning and artificial intelligence.[57] (Canny companies, from MUBI to banks, take advantage of this distrust and advertise on the basis of their human touch, rather than soulless algorithms: see Figure 9.) As ever with new-media talk, these dystopian scenarios are met by optimistic prophecies of technological potential. These include wild promises about automation-induced workers' paradises. In utopian scenarios, think tanks foresee four-day work weeks and guaranteed basic income for human laborers as fruits of efficiency and productivity.[58]

Arts and culture are clearly not immune. Algorithms already compose newspaper stories, including much local reporting and sports coverage; the *Guardian* counts among the clients of these bots.[59] In 2018 the first auction of an algorithmically created painting, *Portrait of Edmond Belamy*, took place. The artwork was generated by feeding a computer information about fifteen thousand paintings made between the fourteenth and twentieth centuries.[60] In 2016, a short algorithm-scripted sci-fi film, *Sunspring*, premiered. We must consider these reports in concert with now long-standing trends toward redundancies and freelance status among film critics, not to mention the casualization of humanities scholars, with adjunct professors and sessional tutors becoming the new norm, another cog in the gig economy.

This book has argued that recommender systems are not endangering human cultural intermediation. But could a more limited thesis nonetheless apply, that these tools might endanger the careers of human cultural intermediators, of film critics, arts and culture journalists, and humanities

FIGURE 9. In January 2018, an advertisement for Jupiter Asset Management was omnipresent on the side of black cabs in London and other UK cities: "Driven by Humans: Using People Not Algorithms to Navigate the Markets." With clever dissembling and counterintuitive branding, a financial services company appealed to a wider popular backlash against the use of algorithms in leisure, entertainment, and artistic culture, indicting their cold rationality and low credibility. Credit: Skully/Alamy Live News.

educators: the automation not of culture but of some aspects of the traditional culture-industry para-professions? Could computers and robots usurp some of the trusted critic's and the learned professor's traditional functions? Will arts criticism become more or less a hobby, a boutique industry of human-touch curation thriving only among the monied classes, patronized by the one-percenters? The philistines who patrol the halls of power would greet the prospect that algorithms could substitute for the subjective pronouncements of ragamuffin humanities scholars (and above all their bodies and salaries) with collaborative filtering and a contract with Netflix U. or the College of the Rotten Tomatoes. This is not entirely science fiction. In 2017, the UK Parliament passed the Higher Education and Research Act, which aimed to allow Google and other providers access to the hitherto almost exclusively state-sponsored higher education market.

It would be foolish to predict definitive answers to these questions. Nevertheless, it is essential to examine with nuance countervailing evidence that suggests that even this more circumscribed proposition may not come to fruition in its most virulent form. Scholars, activists, and other commentators, from the Luddites and John Maynard Keynes to Albert Einstein, have prophesized machine-led takeovers of human labor since at least the early nineteenth century.[61] When faced with the grim statistics that one-quarter of US jobs will suffer "high exposure to automation over the next several decades," economist Paul Krugman wondered in 2019: "When, in modern history, has something like that statement *not* been true?" Recalling the late 1940s automation scares and the subsequent boom in US-American employment, Krugman dismisses arguments that "this time is different" on account of the internet and big-data technological change. The overall sluggish productivity gains—a proxy for measuring automation—between 2007 and 2018 compared to 1995 to 2006 reveal that there is no mass or accelerating machine-learning replacement of human labor underway. Indeed, automation is slowing.[62]

The suspicion—and outright hostility—toward Netflix, Facebook, or Uber has been directed squarely at algorithms; in these critiques, computational processes function as the emblem for the soulless destruction of human jobs. Nevertheless, we should not locate the sins of a company like Uber in its app and GPS algorithm; they replace phoning a human dispatcher in a way that potentially benefits both drivers and passengers, like the online dictionary has replaced time-consuming page-flipping. The algorithm is not necessarily the problem, but rather how Uber has used its technology as an excuse to casualize working contracts and skirt the spirit of existing labor laws. Perhaps algorithms are less culpable than feckless governments and toothless regulators that fail to develop and enforce statutory mechanisms to classify the service as a taxi provider, the company's ultimate use value. On this larger issue I agree with William Uricchio: "Blaming 'the algorithm' or 'big data' puts us in the position of a bull fixated on the matador's cape: we fail to see the real source of malice."[63] I also stand shoulder to shoulder with Tarleton Gillespie, who urges that we "must not conceive of algorithms as abstract, technical achievements, but must unpack the warm human and institutional choices that lie behind these cold mechanisms."[64]

We must not let ourselves be blinded by science: algorithms are just the shiny new tools in a grand old ideological project to privatize and regressively redistribute capital. In this sense, we urgently require regulation (and enforcement) that shores up antitrust statutes and breaks up anti-competitive monoliths like Amazon and Netflix (and Disney and so on). Governments must register and fairly tax revenue where the service occurs, and not in whatever tax-haven-of-the-month these media behemoths have incorporated their shell companies. Strong data protection legislation is essential in a world where personal information and tastes are potentially more valuable than oil, and leaks and hacks run rampant. Technology- and media-literacy projects remain vital so that marketing myths lose their influence on understandings of algorithms and recommender systems. Likewise, in countries with weak traditions of public cultural subsidies such as the United States, initiatives must take hold to support local and national constellations of rigorous film and series commentary and context. And, in the meantime, we need robust quotas for VOD services to carry local and national productions, as well as insurance that recommendation idioms present rather than bury such titles. The European Union and other governmental bodies have made progress on these issues in recent years; as scholars, critics, and activists we should devote our energies to these projects and other positive changes along these lines, rather than succumbing to nothing-to-be-done victim narratives. Techno-deterministic arguments about irrevocable cultural changes because of VOD services' opaque algorithms ultimately enervate and deactivate human agency.

Clearly, I cannot guarantee that well-paid film critics will be (re-)hired in droves. But this has more to do with the diluted ranks of paid jobs in local newspapers, publishers' need to derive a sustaining business model for online journalism, than any culture-changing Netflix algorithm. In this vein, it is altogether premature to declare the death of criticism, of theory, or of the humanities on the basis of Netflix, Amazon, or Google. History shows how criticism has persevered through a series of supposedly existential crises fueled by the introduction of new media. In each of these episodes, critics have earnestly deliberated on their supposed imminent demise (and the concomitant dumbing down and fracturing of the public sphere) in the face of every major new medium or format: specialist cinephile magazines like *Cahiers du Cinéma*; syndicated and televised film reviewing; internet blogs, social media, and aggregators like Rotten Tomatoes; and now the Netflix recommender system. In each of these periods, criticism did not die: the need for gatekeepers, attention focalizers, context, and information persisted. Nevertheless, in response to the new formats, criticism changed. Seen in this historical context, the present juncture portends a similar scenario. Just as so often in the past, we could crouch into defensive postures, cling on to outmoded norms, or refuse to engage with emerging trends. This is a common response to new media, such as how 1990s newspapers reacted to the widespread adoption of the internet with a reactive, short-termist, and ultimately self-destructive defensiveness.[65]

My advice, however, is that humanists and film critics need to *let go* of some formats and processes that have little intrinsic benefit to either the producers or consumers of cultural recommendations. Even if with regret and nostalgia, we need to relinquish certain low-utility objectives, just like robots take up automated tasks in car assembly or how looking up a word in a computer-powered dictionary is much more efficient than flipping by hand through a paper copy with terms arranged in alphabetical order. To be sure, we might sacrifice some positive externalities of otherwise inefficient processes: I lose the serendipity of stumbling over an interesting word in the process of searching for another. But in general, companies use robots for manufacturing and humans stopped buying paper dictionaries because the new tools represented a net positive tradeoff. In the same way, we should not mourn the more banal aspects and functions of film criticism—the sort that a VOD algorithmic recommender system offers—any more than we mourn MovieFone or having to buy paper-based TV guides and Leonard Maltin's compendia, rather than being able to access Rotten Tomatoes, Metacritic, on-screen listings, or a review or forum online.

I have serious reservations about Netflix's and Amazon's claims to democratize access, grave doubts about whether VOD recommender systems can best tips from a seventeen-year-old video shop clerk. Nevertheless, there are situational user purposes, cultural interactions, and desires ("easy-watching"; "I want to be engaged for the next two hours after work") that do not require learned expertise, challenging stylistic subversion, or fusillades of provocation, defamiliarization, and self-reflexivity. Culture, and cultural needs, are often—yes, most of the time and for most users—banal. Diversity is essential; aesthetic innovation is vital. But we must not allow ourselves to become tribalistic in our ways of seeing and thinking or begin to erect border walls around whole swaths of human activities and needs. Very few humanities scholars these days write books by hand or on a mechanical typewriter; as humanists we should likewise advance a confident understanding of our vocation and self-worth, one that lays claim to ambitious research questions but avoids methodological territorialism. Human judgments are not better simply because they are human any more than technology is good (or bad) because it is technological. Algorithms do not "rule our world" any more than paper, printing presses, or ink ruled our world in previous eras or still do so now.

The introduction of VOD recommenders—of various technologies, interfaces, content catalogs, informational features, and designs—offer at least the potential to free up film criticism for higher purposes. Media scholars have long reminded us of how the introduction of one new medium does not make superfluous but usually changes the purpose of the legacy medium: photography and cinema liberated painting from representational tasks, the introduction of television changed radio from a storytelling medium to one focused on music, news, and commentary, and so on.[66]

Higher vistas remain for criticism. It is clear from this study that users still hunger for criticism's third-stage purpose: to deepen engagement *after* the film, to test opinions, to enter into an imagined dialogue about cultural value. But quick-tip listings, an important informational service that critics have long provided, may well yield further to aggregators and algorithms. In this, the distinction between "reviewing" and "criticism"—for most of history a critic-propagated false dichotomy (rather than a long and subtle spectrum) of "consumer guide" versus "guardians of culture and canons"—may in fact become sharper as traditional reviewing makes way for VOD recommender systems and the "like" economy, on the one hand, and multimedia curatorial practices and long-form criticism, on the other.[67] Some professions will change; some professionals may lose their livelihood, or at least the livelihood they imagined. It is hardly a zero-sum game; however, there will be both winners and losers in this transition.

This book has illuminated cultural taste as a contested space, an active site of partisan debate. Technological intrusions cause anxiety and evoke utopian hopes for better recommendation—more accurate, more objective, more democratic, financially lucrative. This is a story of changing filters and gatekeepers, automation and agency, the first chapter in a future history of how VOD services and their recommender systems become ingrained habits, second-nature social protocols. The "success of all media," Lisa Gitelman writes, "depends at some level on inattention or 'blindness' to the media technologies themselves (and all of their supporting protocols) in favor of attention to the phenomena, 'the content,' that they represent for users' edification or enjoyment."[68] Nevertheless, we can reflect, just a bit, on our own consumption habits and from where they derive. Whether via considered deliberation, "heavy curation," or distracted advertisement- or algorithm-led presentation, each of our choices, our taste, has a constellation of origins.

Throughout history, a common pattern coincides with the introduction of a new medium, whether the telephone, cinema, television, the internet, or VOD recommender systems. There are claims that this new medium will revolutionize users' lives and will rebalance the hierarchy between elites and commoners, producers and consumers, city and countryside, and so on. This thinking begins with the marketing rhetoric from those who stand to benefit financially from the new technology and then infects the para-industry: *Wired* magazine and its precursors and cognates, newspaper reporters and columnists, and academia. In nearly every case, however, the story ends as follows: first, the new medium is not as revolutionary, democratic, or anti-democratic as advertised; second, large companies or corporations succeed in controlling and monetizing the new medium for commercial gain; and, third, the new medium sets in motion long-term sociocultural changes different from those initially predicted.

This book has sought to challenge the exceptionalism of algorithmic recommender systems, VOD, the internet, and big data, and to show how supposedly novel modes of cultural recommendation remain indebted to the past. Despite a variety of technological forms and business models, some basic facts endure: the quest for credibility and authority of recommendation, debates about who has the right to determine issues of taste, questions about whether cultural suggestions should confirm preexisting preferences or should elevate, and disagreements over whether taste is exogenous, preformed, and innate to individuals or groups before their encounters with media systems, or remains endogenous and able to be shaped by media and society. The "new" recommender systems are not unique, or inevitable, or without alternatives. Although we have settled on these solutions for now, their forms are provisional even as their underlying commercial mandates and consumer desires persist.

In this sense, it may be less important to be suspicious of *new media*; rather, we need to cultivate a productive suspicion of the *new* as label and discourse. Yes, online, data-fueled recommender systems may seem disruptive and unprecedented. At their core, however, they merely resuscitate long-standing debates over the commodification of taste and how cultural preferences should be organized, policed, guided, and led—discourses that far precede the internet. More than just a *plus ça change* plaidoyer, the larger stakes of this book gesture toward examinations of the "new" via the old, the normal, the used, embodied experience, and the everyday. Toward histories and archaeologies of whinging and fantasizing. Toward a differentiation between originality and novelty.

APPENDIX

Designing the Empirical Audience Study

SURVEY QUESTIONS AND RESPONSES

Because of my research questions, I envisioned a mixed-method design from the start. On the one hand, in order to be able to test, challenge, or advance knowledge from the many theoretical pronouncements that had been made on the subject, I needed the representative results that a quantitative survey could provide. It was also important to launch my survey in both the United Kingdom and United States as a way to provide a possible comparison between two similarly wealthy and English-speaking countries that nonetheless possessed different media landscapes (e.g., the United Kingdom's strong public service broadcast tradition) and different speeds and shapes of VOD adoption (e.g., Netflix streaming began slowly in the United States in 2007, whereas the company first launched in the United Kingdom at the beginning of 2012). On the other hand, the qualitative part of the study was intended to corroborate and illustrate the quantitative measures, to deepen and develop the findings with elaboration and clarification, as well as to address my explicitly inductive research questions about how real users speak about their recommender system use in the context of other information sources.

Survey questions and answer options were designed with the European Union's 2014 study *A Profile of Current and Future Audience* in mind.[1] In most cases I used the same response formulations and explanations—for example, "freshness (titles added most recently to catalogue)"—as a way to be able to confirm, refute, or chart changes between that older study, whose fieldwork had been carried out in 2013 and thus only in the midst of the widespread introduction of streaming in Europe. In a handful of instances, I consciously used a variant formulation, omitted a response option, or added a response option. For example, the EU study did not include "critics' reviews" or "the score on review aggregators (Rotten Tomatoes, Metacritic, etc.)" as a possible response for VOD viewing (whereas my survey did do). In contrast, the EU included the possibility to select "prior knowledge (film

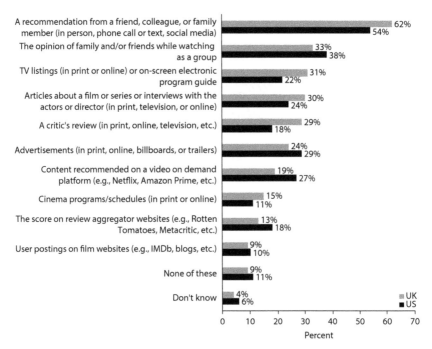

A recommendation from a friend, colleague, or family member (in person, phone call or text, social media) — 62% / 54%

The opinion of family and/or friends while watching as a group — 33% / 38%

TV listings (in print or online) or on-screen electronic program guide — 31% / 22%

Articles about a film or series or interviews with the actors or director (in print, television, or online) — 30% / 24%

A critic's review (in print, online, television, etc.) — 29% / 18%

Advertisements (in print, online, billboards, or trailers) — 24% / 29%

Content recommended on a video on demand platform (e.g., Netflix, Amazon Prime, etc.) — 19% / 27%

Cinema programs/schedules (in print or online) — 15% / 11%

The score on review aggregator websites (e.g., Rotten Tomatoes, Metacritic, etc.) — 13% / 18%

User postings on film websites (e.g., IMDb, blogs, etc.) — 9% / 10%

None of these — 9% / 11%

Don't know — 4% / 6%

■ UK ■ US

FIGURE 10. Question 1. For the following question, please imagine that you are choosing a film or series to watch (e.g., in a cinema, on television, on DVD, online, etc.). Please do NOT think about individual TV programs (e.g., news or sports programs that are not part of a series). In general, which, if any, of the following are LIKELY to influence your choice of what to watch? (Please select all that apply.)

Unweighted base (for UK survey): All UK adults online (n = 2,123)

Unweighted base (for US survey): All US adults online (n = 1,300)

I have heard about)." I found this latter response to unhelpfully conflate several possible sources and thus left it out from my study.

The quantitative surveys were conducted by YouGov Plc and carried out online. Total sample sizes were 2,123 UK adults and 1,300 US adults. The fieldwork was undertaken between 13 and 14 November 2018 in the United Kingdom, and between 13 and 15 November 2018 in the United States. The figures have been weighted and are representative of all UK and US adults (aged 18 and over), respectively.

INTERVIEW SAMPLING

The interviews took place between November 2018 and September 2019 and were conducted by the project research assistant or by the author. Interviewees were ini-

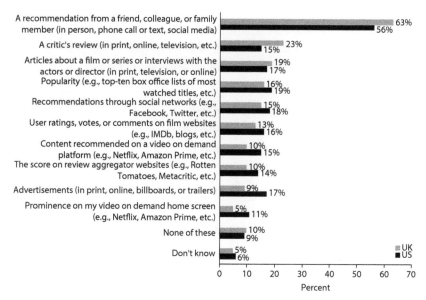

FIGURE 11. Question 2. Still imagining that you are choosing a film or series to watch (e.g., in a cinema, on television, on DVD, online, etc.). Which, if any, of the following forms of film and series recommendation are you MOST LIKELY to trust? (Choose up to THREE.)

Unweighted base (for UK survey): All UK adults online ($n = 2,123$)

Unweighted base (for US survey): All US adults online ($n = 1,300$)

tially sought out using a snowball sampling method but always with the intention to use a sequential approach: to gradually add interviewees in an evolving iterative process of theoretical reflection and refinement. Squarely in the tradition of qualitative audience research, the aim was not to produce a representative sample of VOD users (or film and series users overall) so as to generate generalizable results. Nevertheless, in the course of soliciting interviewees, care was taken to purposively include a sufficient number of participants (and/or achieve some balance) in terms of gender, age, race, heavy and light users (of VOD, but also of film and series consumption overall), region of the country, urban/suburban/rural, as well as education level. This meant that, although the bulk of the interviews took place within a few months, soliciting the last remaining interviewees of a few relatively hard-to-reach groups (e.g., aged in fifties or sixties, living in rural areas, non-university educational backgrounds) took considerable effort and time. The iterative process to theoretically reflect on the initial set of data and then collect further data, however, was necessary to ensure sufficient variety in the sample members—in other words, to have interviewees who offered differing characteristics relevant to the research questions—and produce more robust results.

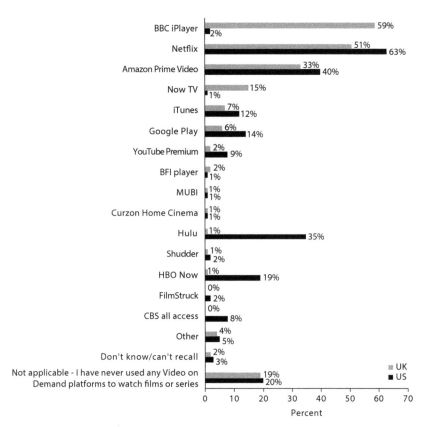

FIGURE 12. Question 3. Which, if any, of the following Video on Demand platforms have you ever used to watch films or series? (Please select all that apply. If you have never used any Video on Demand platforms, please select the "not applicable" option.)

Unweighted base (for UK survey): All UK adults online ($n = 2{,}123$)

Unweighted base (for US survey): All US adults online ($n = 1{,}300$)

Interviews were audio-recorded, transcribed, read, and coded using NVivo software. Already after approximately fifteen interviews, it was clear that the difference between heavy and light users (and between the then emerging typologies of high-stakes and low-stakes use) were among the most important and decisive criteria for information-source behaviors. After twenty-six interviews, I felt I was approach-

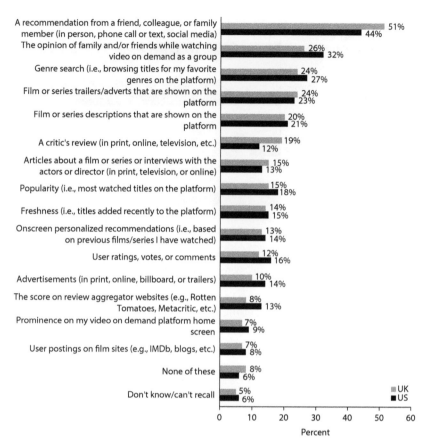

FIGURE 13. Question 4. You previously said that you have used at least one Video on Demand platform to watch films or series. Thinking about when you have chosen a film or series to watch on a VIDEO ON DEMAND platform . . . Which, if any, of the following are MOST LIKELY to influence your choice of what to watch? (Please select all that apply.)

Unweighted base (for UK survey): All UK adults online who have used VOD platforms to watch films or series ($n = 1,692$).

Unweighted base (for US survey): All US adults online who have used VOD platforms to watch films or series ($n = 1,042$).

ing data saturation, if not yet theoretical saturation.[2] We conducted nine further interviews, especially to achieve a richer demographic spread and to ensure that we could capture a full range of perspectives to answer the final, inductively approached research question on how real users speak about their recommender systems use in the context of other information sources.

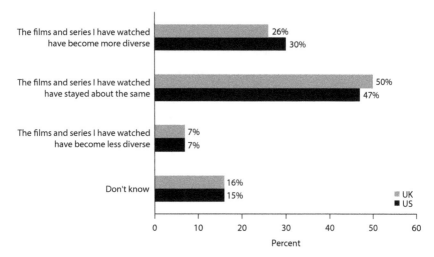

FIGURE 14. Question 5. For the following question, by "diverse" we mean films or series offered in languages that you do not usually watch, made from a different time period, or outside of your usual/favorite genres. Overall, would you say that the films and series you have watched on Video on Demand platforms are more or less diverse compared to before you started using these platforms, or would you say they are about the same?

Unweighted base (for UK survey): All UK adults online who have used VOD platforms to watch films or series ($n = 1,692$).

Unweighted base (for US survey): All US adults online who have used VOD platforms to watch films or series ($n = 1,042$).

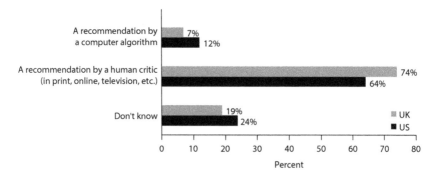

FIGURE 15. Question 6. Now thinking about watching a film or series on ANY platform (i.e., NOT limited to Video on Demand) . . . If you had to choose, which type of recommendation are you more likely to trust? (Please select the option that BEST applies.)

Unweighted base (for UK result): All UK adults ($n = 2,123$)

Unweighted base (for US result): All US adults ($n = 1,300$)

TABLE 2 Interview Participants

#1	male	early 30s	white	SW England
#2	female	late 20s	BAME	Wales
#3	male	late 30s	white	Wales
#4	male	early 30s	BAME	Northern England
#5	female	early 30s	white	Northern England
#6	male	early 30s	BAME	London
#7	female	late 30s	BAME	Northern England
#8	female	early 30s	white	Wales
#9	female	late 40s	BAME	Wales
#10	male	late teens	BAME	Wales
#11	male	late 20s	BAME	Northern England
#12	male	early 40s	white	Wales
#13	male	late 30s	white	Wales
#14	female	early 30s	BAME	Wales
#15	female	early 30s	BAME	Wales
#16	female	early 30s	white	Wales
#17	female	late 20s	white	Wales
#18	female	late 20s	white	Wales
#19	female	late 20s	white	Wales
#20	female	late 20s	white	Wales
#21	male	late 20s	white	Scotland
#22	female	late 20s	white	Wales
#23	female	late 20s	white	SW England
#24	male	late 20s	white	Wales
#25	male	early 20s	white	Wales
#26	female	early 50s	white	Wales
#27	male	early 50s	white	Wales
#28	female	late 50s	white	Wales
#29	male	late 50s	white	SE England
#30	female	early 60s	white	Wales
#31	female	early 20s	BAME	SE England
#32	female	early 20s	white	SE England
#33	male	late 20s	white	SE England
#34	female	late 30s	BAME	London

SEMI-STRUCTURED INTERVIEW QUESTION SCHEDULE

Some interviewees filled out a questionnaire beforehand that included basic demographic data, the first five questions from the YouGov survey, plus multiple-choice questions that gauged the interviewee's frequency of watching films or series on television, on DVD/Blu-ray, on VOD, and at the cinema, respectively. Where a pre-interview questionnaire was not feasible, these questions were incorporated into the interview itself. Where possible, interviewees were asked to demonstrate how they typically choose on VOD by logging into their most frequently used streaming service and narrating how they interact with the platform and its information sources. (In some cases, this demonstration was not possible because the interviewee no longer subscribed to a VOD service, or for other logistical reasons.) The interviews were designed as semi-structured so that the interviewer could freely follow up on participants' responses. Interview lengths ranged from 25 to 50 minutes and averaged approximately 40 minutes. The starting-point questions were as follows:

- Tell me about the last film you saw in a cinema. How did you choose the content? What sources of information (listings, reviews, review aggregators, advertisements, a tip from a friend or family member, social media, interviews, etc.) did you use to make your decision?

- Tell me about the last film or series you saw on television. How did you choose the content? What sources of information (electronic program guide, listings, reviews, review aggregators, advertisements, word of mouth, social media, interviews, etc.) did you use to make your decision?

- Tell me about the last film or series you saw on DVD/Blu-ray. How did you choose the content? What sources of information did you use to make your decision?

- Tell me about the last film or series you saw on a VOD service. How did you choose the content? What sources of information (search, genre browse, "recommended for you," reviews, review aggregators, advertisements, word of mouth, social media, interviews, etc.) did you use to make your decision?

- What sorts of information do you most often use? *Or: if working from the form that the interviewee has provided:* I see you often consult (*methods they say they use often, e.g., word of mouth, critics, recommender system*). Why do you use this method? What makes it more trustworthy than other sources? Why don't you use more (*methods they say they use less often*)?

- What are your favorite genres to watch? Are there genres that you don't like to or won't watch?

- Which VOD service(s) do you use most often? Why?

- Please log on to your preferred VOD service and show me how you usually decide what to watch.

- Since you began watching VOD, would you say that the types of films and series you watch have become more or less diverse? (*"Diverse" = films or series in languages other than your native language; films or series made before 1990; films and series outside of your usual favorite genres*).

- VOD services like Netflix and Amazon promise to provide you with personalized recommendations by using algorithms. They argue that human critics provide poor recommendations because their reviews are broadcast and thus non-personalized. Other people say that humans, such as critics, are basing their recommendations on a lifetime of knowledge and so provide better suggestions. Based on your experiences, what do you think about this?

NOTES

INTRODUCTION

1. See, e.g., A. O. Scott, "The Shape of Cinema, Transformed at the Click of a Mouse," *New York Times*, 18 March 2007, http://www.nytimes.com/2007/03/18 /movies/18scot.html; Michael Gubbins, "Digital Revolution: Active Audiences and Fragmented Consumption," in *Digital Disruption: Cinema Moves On-line*, ed. Dina Iordanova and Stuart Cunningham (St. Andrews, UK: St. Andrews Film Studies, 2012), 67–100.

2. See, e.g., Henry Jenkins, *Convergence Culture: Where Old and New Media Collide*, 2nd rev. ed. (New York: New York University Press, 2006).

3. See, e.g., Chris Anderson, *The Longer Long Tail: How Endless Choice Is Creating Unlimited Demand*, 2nd rev. ed. (London: Random House, 2009).

4. See, for just one example of this regard, Gubbins, "Digital Revolution," esp. 67–69.

5. Gubbins, "Digital Revolution," 86. See also Jenkins, *Convergence Culture*.

6. Pedro Domingos, *The Master Algorithm: How the Quest for the Ultimate Learning Machine Will Remake Our World* (New York: Basic Books, 2015); James Surowiecki, *The Wisdom of Crowds: Why the Many Are Smarter Than the Few* (London: Abacus, 2005); Joe Weinman, *Digital Disciplines: Attaining Market Leadership via the Cloud, Big Data, Social, Mobile, and the Internet of Things* (Hoboken, NJ: Wiley, 2015).

7. Clive Thompson, "If You Liked This, You're Sure to Love That," *New York Times*, 21 November 2008, https://www.nytimes.com/2008/11/23/magazine /23Netflix-t.html.

8. See, e.g., John Cheney-Lippold, *We Are Data: Algorithms and the Making of Our Digital Selves* (New York: New York University Press, 2017); Eli Pariser, *The Filter Bubble: What the Internet Is Hiding from You* (London: Penguin, 2011); Joseph Turow, *The Daily You: How the New Advertising Industry Is Defining Your Identity and Your Worth* (New Haven, CT: Yale University Press, 2011); Todd Gitlin, "Public Sphere or Public Sphericles?," in *Media, Ritual, Identity*, ed. Tamar Liebes and

James Curran (London: Routledge, 1998), 168–75; Cass R. Sunstein, *Republic.com 2.0* (Princeton, NJ: Princeton University Press, 2007); David Beer, "Algorithms: Shaping Tastes and Manipulating the Circulations of Popular Culture," in *Popular Culture and New Media: The Politics of Circulation* (Basingstoke, UK: Palgrave Macmillan, 2013), 64–98; Christopher Steiner, *Automate This: How Algorithms Took Over Our Markets, Our Jobs, and the World* (New York: Portfolio/Penguin, 2012); Nick Bostrom, *Superintelligence: Paths, Dangers, Strategies* (Oxford: Oxford University Press, 2014); Blake Hallinan and Ted Striphas, "Recommended for You: The Netflix Prize and the Production of Algorithmic Culture," *New Media and Society* 18, no. 1 (2014): 117–37; Ted Striphas, "Algorithmic Culture," *European Journal of Cultural Studies* 18, nos. 4–5 (2015): 395–412; Paul Dourish, "Algorithms and Their Others: Algorithmic Culture in Context," *Big Data and Society* 3, no. 2 (2016): 1–11; Neta Alexander, "Catered to Your Future Self: Netflix's 'Predictive Personalization' and the Mathematization of Taste," in *The Netflix Effect: Technology and Entertainment in the 21st Century*, ed. Kevin McDonald and Daniel Smith-Rowsey (New York: Bloomsbury, 2016), 81–97.

9. See, e.g., Andrew Keen, *The Cult of the Amateur: How Today's Internet Is Killing Our Culture and Assaulting Our Economy* (London: Nicholas Brealey, 2007); Gubbins, "Digital Revolution," 67–100; Striphas, "Algorithmic Culture," 395–412.

10. Vincent Mosco, *The Digital Sublime: Myth, Power, and Cyberspace* (Cambridge, MA: MIT Press, 2004).

11. In their history of online screen distribution since the first online VOD platform IFilm launched in 1997, Stuart Cunningham and Jon Silver posit that compared with "the music business or newspapers and book publishing, new technologies have had a slower impact in screen distribution." They chart stages of development in the business. First, an initial "fragmentation stage" (1997–2002), when barriers to entry were especially low and small, pioneering, largely undercapitalized companies competed but largely disappeared. A second, "shakeout stage" followed, during which barriers to entry increased and mergers and acquisitions proliferated; battles for market share ensued, followed by the entry of global players, such as the Hollywood studios' joint venture, Hulu, not to mention Apple and Amazon. Writing in 2013, Cunningham and Silver saw the market on the cusp of a long third "mature stage," by which an oligopoly of firms dominate the market on the basis of potentially stable profitability. See Stuart Cunningham and Jon Silver, *Screen Distribution and the New King Kongs of the Online World* (Basingstoke, UK: Palgrave Macmillan, 2013), 2–4, 15–16; quote on 2.

12. For an explanation, see Elissa Nelson, "Windows into the Digital World: Distributor Strategies and Consumer Choice in an Era of Connected Viewing," in *Connected Viewing: Selling, Streaming, and Sharing Media in the Digital Era*, ed. Jennifer Holt and Kevin Sanson (New York: Routledge, 2014), 67.

13. See, for example, OfCom, *OfCom Communications Market Report 2017* (London: OfCom, 2017), 41.

14. See Elizabeth Evans and Paul McDonald, "Online Distribution of Film and Television in the UK: Behavior, Taste, and Value," in *Connected Viewing: Selling,*

Streaming, and Sharing Media in the Digital Era, ed. Jennifer Holt and Kevin Sanson (New York: Routledge, 2014), esp. 159–65.

15. For historical origins, see Dietmar Jannach et al., *Recommender Systems: An Introduction* (Cambridge: Cambridge University Press, 2011), esp. ix–xv, as well as the essays in Francesco Ricci, Lior Rokach, and Bracha Shapira, eds., *Recommender Systems Handbook*, 2nd rev. ed. (New York: Springer, 2015).

16. See, for example, the definitions in Jannach et al., *Recommender Systems*; Ricci, Rokach, and Shapira, *Recommender Systems Handbook*; or Meng-Hui Chen, Chin-Hung Teng, and Pei-Chann Chang, "Applying Artificial Immune Systems to Collaborative Filtering for Movie Recommendation," *Advanced Engineering Informatics* 29, no. 4 (2015): 830: "Personalized recommender systems obtain useful information from historical data, such as a user's interests and purchasing behavior, in order to recommend relevant information or products to that user." The quotation in the main text also derives from this source.

17. Digital-media market analyst Matthew Bell, cited in "The Television Will Be Revolutionised," *Economist*, 30 June 2018, 19.

18. Carlos A. Gomez-Uribe and Neil Hunt, "The Netflix Recommender System: Algorithms, Business Value, and Innovation," *ACM Transactions on Management Information Systems* 6, no. 4, article 13 (2015): 1–19; Jane Martinson, "Netflix's Ted Sarandos: 'We Like Giving Great Storytellers Big Canvases,'" *Guardian*, 15 March 2015, https://www.theguardian.com/media/2015/mar/15/netflix-ted-sarandos-house-of-cards. See also the comments of "Ted Sarandos, Chief Content Officer, Netflix," in *Distribution Revolution: Conversations about the Digital Future of Film and Television*, ed. Michael Curtin, Jennifer Holt, and Kevin Sanson (Oakland: University of California Press, 2014), 144: "Algorithms drive our entire website—there isn't an inch of uncalculated editorial space." For a business valuation and claims of user take-up, see Gomez-Uribe and Hunt, "The Netflix Recommender System," 5, 7.

19. Nicholas Negroponte, *Being Digital*, 2nd rev. ed. (New York: Vintage, 1996).

20. Jonah Berger, *Contagious: How to Build Word of Mouth in the Digital Age* (London: Simon and Schuster, 2013), 64; Emanuel Rosen, *The Anatomy of Buzz Revisited: Real-Life Lessons in Word-of-Mouth Marketing* (New York: Doubleday, 2009), 245.

21. See, for example, Greg Elmer, *Profiling Machines: Mapping the Personal Information Economy* (Cambridge, MA: MIT Press, 2004), esp. 4–5, 40–41, 48–50; Pariser, *The Filter Bubble*; Tarleton Gillespie, "The Relevance of Algorithms," in *Media Technologies: Essays on Communication, Materiality, and Society*, ed. Tarleton Gillespie, Pablo J. Boczkowski, and Kirsten A. Foot (Cambridge, MA: MIT Press, 2014), 167–93; Hallinan and Striphas, "Recommended for You," 117–37; Striphas, "Algorithmic Culture," 395–412; Alexander, "Catered to Your Future Self," 81–97; Cheney-Lippold, *We Are Data*; William Urrichio, "Data, Culture and the Ambivalence of Algorithms," in *The Datafied Society: Studying Culture through Data*, ed. Mirko Tobias Schäfer and Karin van Es (Amsterdam: Amsterdam University Press, 2017), 125–37.

22. See, for examples, of this approach, Lisa Gitelman, *Always Already New: Media, History, and the Data of Culture* (Cambridge, MA: MIT Press, 2006); Erkki Huhtamo and Jussi Parikka, eds., *Media Archaeology: Approaches, Applications, and Implications* (Berkeley: University of California Press, 2011).

23. The quotations come from Francesco Ricci, Lior Rokach, and Bracha Shapira, "Recommender Systems Handbook: Introduction and Challenges," in *Recommender Systems Handbook*, 2nd rev. ed., ed. Francesco Ricci, Lior Rokach, and Bracha Shapira (New York: Springer, 2015), 2. For historical origins, see this book or Jannach et al., *Recommender Systems*. For an example of the early computer science literature that explicitly drew on legacy recommendation forms as models, see Upendra Shardanand and Pattie Maes, "Social Information Filtering: Algorithms for Automating 'Word of Mouth,'" in *Proceedings of the SIGCHI Conference on Human Factors in Computer Systems*, ed. Irvin R. Katz et al. (New York: ACM, 1995), 210–17.

24. Bezos: "I want to transport online bookselling . . . back to the days of the small bookseller, who got to know you very well and would say things like, 'I know you like John Irving, and guess what, here's this new author, I think he's a lot like John Irving.'" Quoted in Robert Spector, *Amazon.com: Get Big Fast* (New York: HarperBusiness, 2000), 142.

25. See Jean-Luc Chabert, ed., *A History of Algorithms: From the Pebble to the Microchip*, trans. Chris Weeks (Berlin: Springer, 1999); Lev Manovich, "Can We Think without Categories?," *Digital Culture and Society* 4, no. 1 (2018): 17–28.

26. For a discussion of the fundamental distinctions between SVOD, AVOD, and TVOD/EST, see European Audiovisual Observatory, *VOD Distribution and the Role of Aggregators* (Strasbourg: European Audiovisual Observatory, 2017), 10. See also the lucid discussions of these issues in Amanda D. Lotz, *Portals: A Treatise on Internet-Distributed Television* (Ann Arbor, MI: Maize Books, 2017), 6–14; and Ramon Lobato, *Netflix Nations: The Geography of Digital Distribution* (New York: New York University Press, 2019), 7–11. For a recent conceptualization of VOD from a television studies perspective, see Catherine Johnson, *Online TV* (London: Routledge, 2019).

27. For more on critical media industries studies, see Jennifer Holt and Alisa Perren, eds., *Media Industries: History, Theory, and Method* (Chichester, UK: Wiley-Blackwell, 2009); Timothy Havens, Amanda D. Lotz, and Serra Tinic, "Critical Media Industry Studies: A Research Approach," *Communication, Culture and Critique* 2, no. 2 (2009): 234–53; Matthew Freeman, *Industrial Approaches to Media: A Methodological Gateway to Industry Studies* (London: Palgrave Macmillan, 2016).

28. See Brian Winston, *Media, Technology, and Society: A History: From the Telegraph to the Internet* (London: Routledge, 1998); Gitelman, *Always Already New.*

29. Gitelman, *Always Already New.*

30. Geert Lovink delivered a key early analysis of internet criticism and coupled it with a broader understanding of critical theory and the history of criticism. Nevertheless, he explicitly disavows any intention "to perform a discourse analysis of the dominant techno-libertarian Internet agenda." See Geert Lovink, *My First Recession: Critical Internet Culture in Transition* (Rotterdam: V2_/NAi, 2003), 25.

For more on "vapor theory," see Geert Lovink, "Enemy of Nostalgia: Victim of the Present, Critic of the Future: Peter Lunenfeld Interviewed by Geert Lovink," *PAJ: A Journal of Performance and Art* 24, no. 1 (2002): 5–15, esp. 8. See also Michael Wolff, *Television Is the New Television: The Unexpected Triumph of Old Media in the Digital Age* (New York: Portfolio/Penguin, 2015), esp. 10–12.

31. Mosco, *The Digital Sublime*, 1. The page numbers in this and the next paragraph refer to Mosco's book. For more on the "digital mythology" rhetoric, see Martin Hirst and John Harrison, *Communication and New Media: From Broadcast to Narrowcast* (Oxford: Oxford University Press, 2007), 215–20. As John McMurria details, cable television was also expected to provide solutions to "racial discrimination and 'civil disorder'" in the United States. See his *Republic on the Wire: Cable Television, Pluralism, and the Politics of New Technologies, 1948–1984* (New Brunswick, NJ: Rutgers University Press, 2017), 33.

32. Mosco, *The Digital Sublime*, 28.

CHAPTER I. WHY WE NEED FILM
AND SERIES SUGGESTIONS

1. See, for example, Paul M. Hirsch, "Processing Fads and Fashions: An Organization-Set Analysis of Cultural Industry Systems," *American Journal of Sociology* 77, no. 4 (1972): esp. 651–52.

2. See Phillip Nelson, "Information and Consumer Behavior," *Journal of Political Economy* 78, no. 2 (1970): 311–29. Recent scholarship insists on the blurriness of these categories, seeing cars as also partly experience goods. See Michael Hutter, "Experience Goods," in *A Handbook of Cultural Economics*, 2nd rev. ed., ed. Ruth Towse (Cheltenham, UK: Edward Elgar, 2011), 212. For another general assessment of the characteristics of cultural experience goods, see Michael Hutter, "Information Goods," in *A Handbook of Cultural Economics*, ed. Ruth Towse (Cheltenham: Edward Elgar, 2003), 263–68.

3. Vinzenz Hediger, "Die Werbung hat das erste Wort, der Zuschauer das letzte: Filmwerbung und das Problem der symmetrischen Ignoranz," in *Strategisches Management für Film- und Fernsehproduktion: Herausforderungen, Optionen, Kompetenzen*, ed. Michael Hülsmann and Jörn Grapp (Munich: Oldenbourg Verlag, 2009), 540. "Symmetrical ignorance" is Arthur DeVany and W. David Walls's play on economist George Akerlof's famous "asymmetrical ignorance" of used car transactions: compared to the buyer, the seller of a used car has an asymmetrical, superior knowledge of the car's quality.

4. For a recent study of audience motivations for watching films, see European Commission, *A Profile of Current and Future Audiovisual Audience* (Luxembourg: Publications Office of the European Union, 2014), 146.

5. Janet Wasko and Eileen R. Meehan, "Critical Crossroads or Parallel Routes? Political Economy and New Approaches to Studying Media Industries and Cultural Products," *Cinema Journal* 52, no. 3 (2013): 153.

6. On these issues, see Joëlle Farchy, "Die Bedeutung von Information für die Nachfrage nach kulturellen Gütern," trans. Vinzenz Hediger, in *Demnächst in Ihrem Kino: Grundlagen der Filmwerbung und Filmvermarktung*, ed. Vinzenz Hediger and Patrick Vonderau (Marburg: Schüren, 2005), 193–211.

7. Farchy, "Die Bedeutung von Information," 194–95. Emphasis in original; author's translation.

8. See Anindita Chakravarty, Yong Liu, and Tridib Mazumdar, "The Differential Effects of Online Word-of-Mouth and Critics' Reviews on Pre-release Movie Evaluation," *Journal of Interactive Marketing* 24, no. 3 (2010): 185; Farchy, "Die Bedeutung von Information," 199.

9. See Susan M. Mudambi and David Schuff, "What Makes a Helpful Online Review? A Study of Customer Reviews on Amazon.com," *MIS Quarterly* 34, no. 1 (2010): esp. 189. For a controlled experiment on consumers' assessment of professional versus user-generated reviews, see Ruud S. Jacobs et al., "Everyone's a Critic: The Power of Expert and Consumer Reviews to Shape Readers' Post-viewing Motion Picture Evaluations," *Poetics* 52 (2015): 91–103.

10. On critics' impact, see Jehoshua Eliashberg and Steven M. Shugan, "Film Critics: Influencers or Predictors?," *Journal of Marketing* 61, no. 2 (1997): 68–78; Bruce A. Austin, "Rating the Movies," *Journal of Popular Film and Television* 7, no. 4 (1980): 384–99; Robert O. Wyatt and David P. Badger, "How Reviews Affect Interest in and Evaluation of Films," *Journalism Quarterly* 61, no. 4 (1984): 874–78; Suman Basuroy, Subimal Chatterjee, and S. Abraham Ravid, "How Critical Are Critical Reviews? The Box Office Effects of Film Critics, Star Power, and Budgets," *Journal of Marketing* 67, no. 4 (2003): 103–17; David A. Reinstein and Christopher M. Snyder, "The Influence of Expert Reviews on Consumer Demand for Experience Goods: A Case Study of Movie Critics," *Journal of Industrial Economics* 53, no. 1 (2005): 27–51; Morris B. Holbrook, "Popular Appeal versus Expert Judgments of Motion Pictures," *Journal of Consumer Research* 26, no. 2 (1999): 144–55. For a discussion of the persistent overestimation of famous critics' "make-or-break" power, see also Mattias Frey, *The Permanent Crisis of Film Criticism: The Anxiety of Authority* (Amsterdam: Amsterdam University Press, 2015), 101–23. On the decisive role of word of mouth in film consumption decisions, see Arthur S. De Vany and W. David Walls, "Bose-Einstein Dynamics and Adaptive Contracting in the Motion Picture Industry," *Economic Journal* 106, no. 439 (1996): 1493–514.

11. See, in addition to Farchy's overview of 1950s French cinema audience questionnaires: Ronald J. Faber, Thomas C. O'Guinn, and Andrew P. Hardy, "Art Films in the Suburbs: A Comparison of Popular and Art Film Audiences," in *Current Research in Film: Audiences, Economics, and Law,* vol. 4, ed. Bruce A. Austin (Norwood, NJ: Ablex, 1988), 45–53; Chakravarty, Liu, and Mazumdar, "The Differential Effects of Online Word-of-Mouth and Critics' Reviews," 185–97; European Commission, *A Profile of Current and Future Audiovisual Audience.*

12. Farchy, "Die Bedeutung von Information," 204–207. For a typology of film viewers including discrete categories and criteria for light and heavy users,

see European Commission, *A Profile of Current and Future Audiovisual Audience*, esp. 16, 68.

13. See, for example, Faber, Guinn, and Hardy, "Art Films in the Suburbs," 46; Bruce Austin, "Film Attendance: Why College Students Chose to See Their Most Recent Film," *Journal of Popular Film and Television* 9, no. 1 (1982): 43–49; Ronald J. Faber and Thomas C. O'Guinn, "Effect of Media Advertising and Other Sources on Movie Selection," *Journalism Quarterly* 61, no. 2 (1984): 371–77; Avenue ISR, *Arthouse Convergence: 2016 National Audience Study*, http://www.arthouseconvergence .org/wp-content/uploads/2017/04/AHC-2016-Presentation-v.F.pdf; European Commission, *A Profile of Current and Future Audiovisual Audience*.

14. Peter Boatwright, Suman Basuroy, and Wagner A. Kamakura, "Reviewing the Reviewers: The Impact of Individual Film Critics on Box Office Performance," *Quantitative Marketing and Economics* 5, no. 4 (2007): 401–25; Chakravarty, Liu, and Mazumdar, "The Differential Effects of Online Word-of-Mouth and Critics' Reviews"; Robert East, Kathy Hammond, and Malcolm Wright, "The Relative Incidence of Positive and Negative Word of Mouth: A Multi-Category Study," *International Journal of Research and Marketing* 24, no. 2 (2007): 175–84.

15. Barry Schwartz et al., "Maximizing Versus Satisficing: Happiness Is a Matter of Choice," *Journal of Personality and Social Psychology* 83, no. 5 (2002): 1178–197. For these theories' applications in recommender systems, see Dietmar Jannach et al., *Recommender Systems: An Introduction* (Cambridge: Cambridge University Press, 2011), esp. 236, 246.

16. See Farchy, "Die Bedeutung von Information," 196.

17. Francesco Ricci, Lior Rokach, and Bracha Shapira, "Recommender Systems: Introduction and Challenges," in *Recommender Systems Handbook,* 2nd rev. ed., ed. Francesco Ricci, Lior Rokach, and Bracha Shapira (New York: Springer, 2015), 1.

18. On this question, see Finola Kerrigan, *Film Marketing*, 2nd rev. ed. (Abingdon, UK: Routledge, 2017), 104.

19. See Farchy, "Die Bedeutung von Information," 198.

20. Ricci, Rokach, and Shapira, "Recommender Systems," 8.

21. See Mattias Frey, "Critical Questions," in *Film Criticism in the Digital Age*, ed. Mattias Frey and Cecilia Sayad (New Brunswick, NJ: Rutgers University Press, 2015), 1–20.

22. Quote from Eugene W. Anderson, "Consumer Satisfaction and Word of Mouth," *Journal of Services Research* 1, no. 1 (1998): 6. On "electronic word of mouth" or "online word of mouth," see Thorsten Hennig-Thurau et al., "Electronic Word-of-Mouth via Consumer-Opinion Platforms: What Motivates Consumers to Articulate Themselves on the Internet," *Journal of Interactive Marketing* 18, no. 1 (2004): 38–52; Thorsten Hennig-Thurau, Caroline Wiertz, and Fabian Feldhaus, "Does Twitter Matter? The Impact of Microblogging Word of Mouth on Consumers' Adoption of New Movies," *Journal of the Academy of Marketing Science* 43, no. 3 (2015): 375–94; Jo Brown, Amanda J. Broderick, and Nick Lee, "Word of Mouth Communication within Online Communities: Conceptualizing the Online Social Network," *Journal of Interactive Marketing* 21, no. 3 (2007):

2–20. For systematic popular engagements with word of mouth, see Jonah Berger, *Contagious: How to Build Word of Mouth in the Digital Age* (London: Simon and Schuster, 2013); Emanuel Rosen, *The Anatomy of Buzz Revisited: Real-Life Lessons in Word-of-Mouth Marketing* (New York: Doubleday, 2009).

23. See Caroline Beaton, "Why You Can't Really Trust Negative Online Reviews," *New York Times*, 13 June 2018, https://nyti.ms/2JJVily; Taryn Luna, "Social Media Play Big Role in Movies," *Boston Globe*, 13 March 2013, https://www.bostonglobe.com/business/2013/03/12/movies-depend-social-media-support-for-staying-power-box-office/mDRqLV2AaS1xqmLdFVIN5O/story.html.

24. See Daniel J. O'Keefe, *Persuasion: Theory and Research*, 3rd rev. ed. (Los Angeles: Sage, 2016); Kyung-Hyan Yoo, Ulrike Gretzel, and Markus Zanker, "Source Factors in Recommender System Credibility Evaluation," in *Recommender Systems Handbook*, 2nd rev. ed., ed. Francesco Ricci, Lior Rokach, and Bracha Shapira (New York: Springer, 2015), 689–714.

25. Europe here refers to European Union members states only. European Audiovisual Observatory, *Film Production in Europe: Production Volume, Co-production and Worldwide Circulation* (Strasbourg: European Audiovisual Observatory, 2017), 1.

26. Johanna Koljonen, *Nostradamus Screen Visions 2017* (Gothenburg: Gothenburg Film Festival Office, 2017), 7.

27. See, for example, John Koblin, "Netflix Says It Will Spend Up to $8 Billion on Content Next Year," *New York Times*, 16 October 2017, https://www.nytimes.com/2017/10/16/business/media/netflix-earnings.html?_r=0; Ben Munson, "Amazon, Hulu and Netflix to Triple Original Content Spend by 2022: Report," *Fierce Cable*, 30 January 2018, https://www.fiercecable.com/video/amazon-hulu-and-netflix-to-triple-original-content-spend-by-2022-report; Annlee Ellingson, "Amazon on Pace to Surpass Netflix in Original Content Spending," *L.A. Biz*, 27 November 2017, https://www.bizjournals.com/losangeles/news/2017/11/27/amazon-on-pace-to-surpass-netflix-in-original.html; Todd Spangler, "Netflix Spent $12 Billion on Content in 2018; Analysts Expect That to Grow to $15 Billion This Year," *Variety*, 18 January 2019, https://variety.com/2019/digital/news/netflix-content-spending-2019-15-billion-1203112090/; "The Television Will Be Revolutionised," *The Economist*, 30 June 2018, 17–20; John Koblin, "Peak TV Hits a New Peak, with 532 Scripted Shows," *New York Times*, 9 January 2020, https://nyti.ms/2T8amRK.

28. See OfCom, *Communications Market Report 2017* (London: OfCom, 2017); Nielsen, *Video on Demand: How Worldwide Viewing Habits Are Changing in the Evolving Media Landscape* (New York: Nielsen, 2016); Nielsen, *The Nielsen Total Audience Report Q2 2017* (New York: Nielsen, 2017).

29. See Daniel Holloway and Cynthia Littleton, "FX's John Landgraf Sounds Alarm About Potential Netflix 'Monopoly,' Overall Series Growth," *Variety*, 9 August 2016, http://variety.com/2016/tv/news/fxs-john-landgraf-netflixs-massive-programming-output-has-pushed-peak-tv-1201833825/; Koljonen, *Nostradamus Screen Visions 2017*, 15.

30. Netflix figures from: Todd Spangler, "Amazon Prime Has 4 Times Netflix's Movie Lineup, But Size Isn't Everything," *Variety*, 22 April 2016, http://variety .com/2016/digital/news/netflix-amazon-prime-video-movies-tv-comparison -1201759030/. Figures for the "average" SVOD service derive from European Audio-visual Observatory, *World Film Market Trends: Focus 2017* (Strasbourg: European Audiovisual Observatory, 2017), 8.

31. Already in 2014, 50 percent of US homes, and 25 percent of UK homes, had a subscription to both a pay TV *and* VOD service. See Christian Grece, *The SVOD Market in the EU: Developments 2014/2015* (Strasbourg: European Audiovisual Observatory, 2015), 19.

32. Mark Sweney, "Battle Rejoins between TV's Two Biggest Stars: Netflix and Amazon," *Guardian*, 13 October 2018, https://www.theguardian.com/media /2018/oct/13/netflix-amazon-battle-rejoins-tv-biggest-stars-romanoffs-making-a -murderer.

33. See Motion Picture Association of America, *MPAA Theatrical Market Statistics 2016* (Los Angeles, CA: MPAA, 2016), 8, 25; British Film Institute, *BFI Statistical Yearbook 2016* (London: British Film Institute, 2016), 121.

34. Steven Follows, "How Many Film Festivals Are There in the World?," 19 August 2013, https://stephenfollows.com/many-film-festivals-are-in-the-world/. Follows estimated there were approximately two hundred active festivals in the United Kingdom in 2013. According to Catharine Des Forges, director of the UK-based Independent Cinema Office, there were four hundred festivals active in 2017. See Koljonen, *Nostradamus Screen Visions 2017*, 11. A 2011 study put the figure at thirty-five hundred worldwide. See Brian Moeran and Jesper Strandgaard Pedersen, "Introduction," in *Negotiating Values in the Creative Industries: Fairs, Festivals and Competitive Events*, ed. Brian Moeran and Jesper Strandgaard Pedersen (Cambridge: Cambridge University Press, 2011), 4.

35. See, for example, Marijke de Valck, "Convergence, Digitisation and the Future of Film Festivals," in *Digital Disruption: Cinema Moves On-line*, ed. Dina Iordanova and Stuart Cunningham (St. Andrews, UK: St. Andrews Film Studies, 2012), 117–29.

36. Sources based on data from the Office of National Statistics, https:// www.ons.gov.uk/peoplepopulationandcommunity/populationandmigration /populationestimates/timeseries/ukpop/pop; OECD, http://stats.oecd.org/.

37. British Film Institute, *BFI Research and Statistics: Exhibition August 2017* (London: British Film Institute, 2017), 6; British Film Institute, *BFI Statistical Yearbook 2016*, 125; British Film Institute, *BFI Statistical Yearbook 2014* (London: British Film Institute, 2014), 108; British Film Institute, *BFI Statistical Yearbook 2012* (London: British Film Institute, 2012), 97.

38. See the yearly OfCom report, the MPAA report, and the quarterly Nielsen Total Audience reports. In the United Kingdom, live television use dropped by thirty-six minutes (i.e., 14%) between 2010 and 2016 (OfCom, *Communications Market Report 2017*, 13). According to Nielsen, in 2017 US households that have streaming spent roughly 12 percent of total television time viewing streaming

content, with approximately half of that consumption taking place via Netflix. See Rani Molla, "TV Producers Can Now See How Many People Are Watching Netflix Shows," *Recode*, 18 October 2017, https://www.recode.net/2017/10/18/16495130 /netflix-the-defenders-6-1-million-viewers-season-premiere. One UK study demonstrated that, at the beginning of 2006, 98.8 percent of television was watched live; by December 2017, that figure had dropped to 85.1 percent. ("Catch-Up and Live TV Compared," *BARB*, accessed 15 December 2017, http://www.barb.co.uk /viewing-data/catch-up-and-live-tv-compared/.) On the amount of spend on video entertainment in the United States, see, e.g., Brian Fung, "Why Netflix Is Raising Prices Again and What It Could Mean for Cord-Cutting," *Washington Post*, 5 October 2017, https://www.washingtonpost.com/news/the-switch/wp/2017/10 /05/why-netflix-is-raising-prices-again-and-what-it-could-mean-for-cord-cutting/ ?utm_term=.f8d9449dda97.

39. See Richard Broughton, "Forget Cord-Cutting and Focus on Doubling-Up," *Ampere Analysis*, 10 April 2017, https://www.ampereanalysis.com/blog/ffa3397a -31a5-47ce-9ba4-566e9a9a2848; Colin Dixon, "Keep My Customer—Why Consumers Subscribe to, Stay with, Cancel, and Come Back to Online Video Services," *NScreenMedia*, 26 March 2019, https://nscreenmedia.com/keep-my -customer/; Reed Albergotti and Sarah Ellison, "Apple Auditions for Hollywood: The Making of a Streaming Service," *Washington Post*, 22 March 2019, https://www .washingtonpost.com/technology/2019/03/22/apple-auditions-hollywood-making -streaming-service/?noredirect=on.

40. See OfCom, *Communications Market Report 2016* (London: OfCom, 2016), 15–16; cf. OfCom, *Media Nations: UK 2019* (London: OfCom, 2019); Nielsen, *Nielsen Total Audience Report Q2 2017*, 13–14ff.

41. See Peter J. Rentfrow, Lewis R. Goldberg, and Ran Zilca, "Listening, Watching, and Reading: The Structure and Correlates of Entertainment Preferences," *Journal of Personality* 79, no. 2 (2011): 223–58; here 223.

42. Frances Perraudin, "Average Briton Spends 26 Days a Year Watching On-demand TV," *Guardian*, 23 September 2018, https://www.theguardian.com/society /2018/sep/23/average-briton-spends-26-days-a-year-watching-on-demand-tv-survey.

43. James G. Webster, *The Marketplace of Attention: How Audiences Take Shape in a Digital Age* (Cambridge, MA: MIT Press, 2014), 4.

44. See Georg Franck, Ökonomie der Aufmerksamkeit: Ein Entwurf (Munich: Carl Hanser, 1998), esp. 49–74.

45. On decreasing average shot lengths, see James E. Cutting et al., "Quicker, Faster, Darker: Changes in Hollywood Film over 75 Years," *i-Perception* 2, no. 6 (2011): 569–76; on artists' needs to intensify style, see Richard A. Lanham, *The Economics of Attention: Style and Substance in the Age of Information* (Chicago: University of Chicago Press, 2006), esp. 53–54; on the transition to increased sex and violence, see Mattias Frey, *Extreme Cinema: The Transgressive Rhetoric of Today's Art Cinema Culture* (New Brunswick, NJ: Rutgers University Press, 2016) and Robert H. Frank and Philip J. Cook, *The Winner-Take-All Society: Why the Few at the Top Get So Much More Than the Rest of Us*, 2nd rev. ed. (London: Virgin,

2010), 189–209; on transmedia storyworlds, see Sarah Atkinson, *Beyond the Screen: Emerging Cinema and Engaging Audiences* (New York: Bloomsbury, 2014).

46. Franck, *Ökonomie der Aufmerksamkeit*; Lanham, *The Economics of Attention*; Webster, *Marketplace of Attention*.

47. Webster, *Marketplace of Attention*, 1.

48. Quote from Webster, *Marketplace of Attention*, 3. On "bounded rationality" in online information processing, see Miriam J. Metzger, Andrew J. Flanagin, and Ryan B. Medders, "Social and Heuristic Approaches to Credibility Evaluation Online," *Journal of Communication* 60, no. 3 (2010): esp. 416–17. On "bounded rationality" in general decision-making, see Gerd Gigerenzer and Reinhard Selten, eds., *Bounded Rationality: The Adaptive Toolbox* (Cambridge, MA: MIT Press, 2001).

49. Webster, *Marketplace of Attention*, 20–21, 138.

50. See James R. Bettman, Mary Francis Luce, and John W. Payne, "Constructive Consumer Choice Processes," *Journal of Consumer Research* 25, no. 3 (1998): 187–217.

51. Webster, *Marketplace of Attention*, 29–30, 137–38.

52. Webster, *Marketplace of Attention*, 36.

53. On this point, see also Amanda D. Lotz, *The Television Will Be Revolutionized*, 2nd rev. ed. (New York: New York University Press, 2014), 63: "despite exponential growth in availability, the number of channels viewed by a household tended to increase only slightly. A household with 31 to 40 channels viewed an average of 10.2, while those with 51 to 90 viewed just over 15." The average number stagnated at 17 even in the period 2008–2014, when the average channels available had surged to 189.

54. Webster, *Marketplace of Attention*, 14, 36–37. On repertoires, see also Uwe Hasebrink and Hanna Domeyer, "Media Repertoires as Patterns of Behaviour and as Meaningful Practices: A Multimethod Approach to Media Use in Converging Media Environments," *Participations* 9, no. 2 (2012): 757–79.

55. Webster, *Marketplace of Attention*, 14, 37–38. See also Metzger, Flanagin, and Medders, "Social and Heuristic Approaches to Credibility Evaluation Online," 413–39.

CHAPTER 2. HOW RECOMMENDER SYSTEMS WORK

1. Carlos A. Gomez-Uribe and Neil Hunt, "The Netflix Recommender System: Algorithms, Business Value, and Innovation," *ACM Transactions on Management Information Systems* 6, no. 4, article 13 (2015): 7.

2. James G. Webster, *The Marketplace of Attention: How Audiences Take Shape in a Digital Age* (Cambridge, MA: MIT Press, 2014), 67.

3. Webster, *Marketplace of Attention*, 130.

4. For more on these industry-led classifications, see Christian Grece, *The SVOD Market in the EU: Developments 2014/2015* (Strasbourg: European Audiovisual Observatory, 2015), 20–22; Gilles Fontaine and Patrizia Simone, *VOD Distribution*

and the Role of Aggregators (Strasbourg: European Audiovisual Observatory, 2017), 16–17.

5. See Mattias Frey, *MUBI and the Curation Model of Video on Demand* (London: Palgrave Macmillan, forthcoming).

6. See Dietmar Jannach et al., *Recommender Systems: An Introduction* (Cambridge: Cambridge University Press, 2011), 1.

7. See Jannach et al., *Recommender Systems*, ix; Will Hill et al., "Recommending and Evaluating Choices in a Virtual Community of Use," in *Proceedings of the SIGCHI Conference on Human Factors in Computing Systems*, ed. Irvin R. Katz et al. (New York: ACM, 1995): 194–201; Robert Armstrong et al., "WebWatcher: A Learning Apprentice for the World Wide Web," in *Information Gathering from Heterogeneous, Distributed Environments: Papers from the AAAI Spring Symposium*, ed. Craig Knoblock and Alon Levy (Palo Alto, CA: AAAI Press, 1995), 6–12.

8. See, for example, Blake Hallinan and Ted Striphas, "Recommended for You: The Netflix Prize and the Production of Algorithmic Culture," *New Media and Society* 18, no. 1 (2014): 117–37.

9. On the legitimation and increased audience interest in films appearing at festivals or that are nominated for or win awards, see Shyon Baumann, *Hollywood Highbrow: From Entertainment to Art* (Princeton, NJ: Princeton University Press, 2007); John C. Dodds and Morris B. Holbrook, "What's an Oscar Worth? An Empirical Estimation of the Effects of Nominations and Awards on Movie Distribution and Revenues," in *Current Research in Film: Audiences Economics, and Law*, vol. 4, ed. Bruce A. Austin (Norwood, NJ: Ablex, 1988), 72–88; Eva Deuchert, Kossi Adjamah, and Florian Pauly, "For Oscar Glory or Oscar Money? Academy Awards and Movie Success," *Journal of Cultural Economics* 29, no. 3 (2005): 159–76. See also Finola Kerrigan, *Film Marketing*, 2nd rev. ed. (Abingdon, UK: Routledge, 2017), esp. 100–120; Barrie Gunter, *Predicting Success at the Box Office* (London: Palgrave Macmillan, 2018), 71–87, 195–225.

10. On Netflix's internal valuation of its recommender system, see Gomez-Uribe and Hunt, "The Netflix Recommender System," 7. On the Wall Street take on Netflix as a "platform company" rather than a "content company," see Joe Nocera, "Can Netflix Survive in the New World It Created?," *New York Times*, 15 June 2016, https://www.nytimes.com/2016/06/19/magazine/can-netflix-survive-in-the-new -world-it-created.html.

11. Willy Shih and Stephen Kaufman, "Netflix in 2011," *Harvard Business School Case Study*, no. 615-007, 19 August 2014, 6.

12. Michael Wolff, *Television Is the New Television: The Unexpected Triumph of Old Media in the Digital Age* (New York: Portfolio/Penguin, 2015), 32–33.

13. Nicholas Negroponte, *Being Digital*, 2nd rev. ed. (New York: Vintage, 1996), 20.

14. Negroponte, *Being Digital*, 150–51.

15. Negroponte, *Being Digital*, 155.

16. Negroponte, *Being Digital*, 164.

17. Negroponte, *Being Digital*, 4. For exemplary critiques of Negroponte, see Cass R. Sunstein, *Republic.com 2.0* (Princeton, NJ: Princeton University Press, 2007), 38–43, or Joseph Turow, *The Daily You: How the New Advertising Industry Is Defining Your Identity and Your Worth* (New Haven, CT: Yale University Press, 2011), 9.

18. As evidence of the widespread and enduring nature of Negroponte's utopian thinking, see Pedro Domingos, *The Master Algorithm: How the Quest for the Ultimate Learning Machine Will Remake Our World* (New York: Basic Books, 2015), xi–xiv, which narrates a wonderful world of personalized machine learning hidden in the plain sight of daily life: "A learning algorithm is like a master craftsman: every one of its productions is different and exquisitely tailored to the customer's needs" (xiv–xv).

19. See James Surowiecki, *The Wisdom of Crowds: Why the Many Are Smarter Than the Few* (London: Abacus, 2005); Pierre Lévy, *Collective Intelligence: Mankind's Emerging World in Cyberspace* (Cambridge, MA: Perseus Books, 1997).

20. Joe Weinman, *Digital Disciplines: Attaining Market Leadership via the Cloud, Big Data, Social, Mobile, and the Internet of Things* (Hoboken, NJ: Wiley, 2015), 43.

21. Domingos, *The Master Algorithm*, 11.

22. Weinman, *Digital Disciplines*, 176; see also 189.

23. Weinman, *Digital Disciplines*, 189. On this general matter, see also Michael D. Smith and Rahul Telang, *Streaming, Sharing, Stealing: Big Data and the Future of Entertainment* (Cambridge, MA: MIT Press, 2016), 10–11, 75, 128–32.

24. Shih and Kaufman, "Netflix in 2011," 10.

25. Cees van Koppen, Netflix manager public policy EMEA, "Commercial Strategies in the UK TV Market – Audiences, Technology and Discoverability," presentation at the conference "Competition in the UK TV Market: Consumer Trends, Commercial Strategies and Policy Options," London, 24 October 2017.

26. See, for example, Hallinan and Striphas, "Recommended for You," 117–37; Patrick Vonderau, "The Politics of Content Aggregation," *Television and New Media* 16, no. 8 (2014): 717–33; esp. 719.

27. Webster, *The Marketplace of Attention*, 132–33.

28. See, for example, Grant Blank, *Critics, Ratings, and Society: The Sociology of Reviews* (Lanham, MD: Rowman & Littlefield, 2007), 1; Jacqueline Johnson Brown and Peter H. Reingen, "Social Ties and Word-of-Mouth Referral Behavior," *Journal of Consumer Research* 14, no. 3 (1987): 350–62.

29. See Emanuel Rosen, *The Anatomy of Buzz Revisited: Real-Life Lessons in Word-of-Mouth Marketing* (New York: Doubleday, 2009), 11, 130; Jonah Berger, *Contagious: How to Build Word of Mouth in the Digital Age* (London: Simon and Schuster, 2013), 7.

30. Daniel J. O'Keefe, *Persuasion: Theory and Research*, 3rd rev. ed. (Los Angeles: Sage, 2016), esp. xv–xvi, 24, 150–59, 188–94, 207, 216, 253.

31. Anindita Chakravarty, Yong Liu, and Tridib Mazumdar, "The Differential Effects of Online Word-of-Mouth and Critics' Reviews on Pre-release Movie

Evaluation," *Journal of Interactive Marketing* 24, no. 3 (2010): 189. See also Elizabeth J. Wilson and Daniel L. Sherrell, "Source Effects in Communication and Persuasion Research," *Journal of the Academy of Marketing Science* 21, no. 2 (1993): 102–12.

32. See Mattias Frey, *The Permanent Crisis of Film Criticism: The Anxiety of Authority* (Amsterdam: Amsterdam University Press, 2015), esp. 18–20.

33. From Chris Cowan's London Business School TedX talk, "How Word of Mouth Really Works," *YouTube*, 31 May 2016, https://www.youtube.com/watch?v= Z_EERL2hSrg. On this point, see also Rosen, *The Anatomy of Buzz Revisited.* For a literature review on trust, credibility, predictability, and goodwill in film marketing, see Kerrigan, *Film Marketing*, 105–6.

34. See chapter 5 of this book and also European Commission, *A Profile of Current and Future Audiovisual Audience* (Luxembourg: Publications Office of the European Union, 2014), 45–58.

35. Susan M. Mudambi and David Schuff, "What Makes a Helpful Online Review? A Study of Customer Reviews on Amazon.com," *MIS Quarterly* 34, no. 1 (2010): 185–200; esp. 185, 189, 194, 196, 199.

36. Miriam J. Metzger, Andrew J. Flanagin, and Ryan B. Medders, "Social and Heuristic Approaches to Credibility Evaluation Online," *Journal of Communication* 60, no. 3 (2010): 416.

37. Metzger, Flanagin, and Medders, "Social and Heuristic Approaches," 421, 423, 426–30, 432.

38. Jannach et al., *Recommender Systems*, 260–62.

39. Nava Tintarev and Judith Masthoff, "Explaining Recommendations: Design and Evaluation," in *Recommender Systems Handbook,* 2nd rev. ed., ed. Francesco Ricci, Lior Rokach, and Bracha Shapira (New York: Springer, 2015), 370.

40. Jannach et al., *Recommender Systems*, 161.

41. Jannach et al., *Recommender Systems*, 162.

42. Jannach et al., *Recommender Systems*, 163–64. See also Jonathan L. Herlocker, Joseph A. Konstan, and John Riedl, "Explaining Collaborative Filtering Recommendations," *Proceedings of the 2000 ACM Conference on Computer Supported Cooperative Work* (New York: ACM, 2000), 241–50; John O'Donovan and Barry Smyth, "Trust in Recommender Systems," *Proceedings of the 10th International Conference on Intelligent User Interfaces* (New York: ACM, 2005), 167–74.

43. Tarleton Gillespie, "The Relevance of Algorithms," in *Media Technologies: Essays on Communication, Materiality, and Society*, ed. Tarleton Gillespie, Pablo J. Boczkowski, and Kirsten A. Foot (Cambridge, MA: MIT Press, 2014), 179.

44. For a critique of Surowiecki, wisdom of crowds, and collective intelligence, see Webster, *The Marketplace of Attention*, 18, 90–91.

45. Matthew Arnold, *Culture and Anarchy and Other Writings*, ed. Stefan Collini (Cambridge: Cambridge University Press, 1993 [1869]), 190.

46. See, for example, Johann Gottfried Herder, *Selected Writings on Aesthetics*, ed. and trans. Gregory Moore (Princeton, NJ: Princeton University Press, 2006);

David Hume, *Four Dissertations* (London: A. Millar, 1757); Immanuel Kant, *Critique of Judgment*, ed. and trans. J. H. Bernard (London: Macmillan, 1914 [1790]).

47. See Thorstein Veblen, *The Theory of the Leisure Class* (Boston: Houghton Mifflin, 1973 [1899]); Pierre Bourdieu, *Distinction: A Social Critique of the Judgment of Taste*, trans. Richard Nice (London: Routledge, 1984 [1979]).

48. See, for example, Neta Alexander, "Catered to Your Future Self: Netflix's 'Predictive Personalization' and the Mathematization of Taste," in *The Netflix Effect: Technology and Entertainment in the 21st Century*, ed. Kevin McDonald and Daniel Smith-Rowsey (New York: Bloomsbury, 2016), 81–97; Hallinan and Striphas, "Recommended for You."

CHAPTER 3. DEVELOPING NETFLIX'S ALGORITHMS

1. See Jiji Lee, "'What Should I Watch on Netflix?': A New Original Series," *New York Times*, 11 November 2017, https://www.nytimes.com/2017/11/11/opinion /sunday/what-should-i-watch-on-netflix.html.

2. Elsa Keslassy, "Netflix, Cannes Still at Odds over Theatrical Release Rules," *Variety*, 28 November 2018, https://variety.com/2018/film/global/netflix-cannes -film-festival-1203037092/; Michael Schneider, "Feast Full of Dollars," *Variety*, 5 March 2019, 33–37; Brooks Barnes, "Netflix's Movie Blitz Takes Aim at Hollywood's Heart," *New York Times*, 16 December 2018, https://www.nytimes.com /2018/12/16/business/media/netflix-movies-hollywood.html; Caroline Franke, "Easy to Devour," *Variety*, 29 January 2019, 46–49; David Streitfeld, "'*Black Mirror*' Gives Power to the People," *New York Times*, 28 December 2018, https://www .nytimes.com/2018/12/28/arts/television/black-mirror-netflix-interactive.html; Rebecca Nicholson, "Bingewatching: For Me and 40m Others, One Episode Is Never Enough," *Guardian*, 3 August 2017, https://www.theguardian.com/tv-and -radio/2017/aug/03/bingewatching-40m-episode-never-enough-tv-commitment; Jacob Oller, "How Not to Watch TV: Netflix Coins Term 'Binge Race,'" *Forbes*, 17 October 2017, https://www.forbes.com/sites/jacoboller/2017/10/17/how-not-to -watch-tv-netflix-coins-term-binge-race/#319784025128. For scholarly treatments of binge-watching, see Bridget Rubenking and Cheryl Campanella Bracken, *Binge Watching: Motivations and Implications of Our Changing Viewing Behaviors* (Bern: Peter Lang, 2020), as well as the work of Mareike Jenner, e.g., "Binge-Watching: Video-on-Demand, Quality TV and Mainstreaming Fandom," *International Journal of Cultural Studies* 20, no. 3 (2017): 304-20; and her *Netflix and the Re-Invention of Television* (Basingstoke, UK: Palgrave, 2018).

3. OfCom, *Communications Market Report 2017* (London: OfCom, 2017), 21.

4. "About Netflix," https://media.netflix.com/en/about-netflix.

5. Edmund Lee, "Everyone You Know Just Signed Up for Netflix," *New York Times*, 21 April 2020, https://nyti.ms/3cDh1cW; Mark Sweney, "Disney Forecast

to Steal Netflix's Crown as World's Biggest Streaming Firm," *Guardian*, 14 March 2021, https://www.theguardian.com/film/2021/mar/14/disney-forecast-to-steal-netflix-crown-as-worlds-biggest-streaming-firm.

6. Cees van Koppen, Netflix manager of public policy EMEA, "Commercial Strategies in the UK TV Market—Audiences, Technology and Discoverability," presentation at the conference "Competition in the UK TV Market: Consumer Trends, Commercial Strategies and Policy Options," London, 24 October 2017.

7. Hank Stuever, "In Its Bid to Be Everything, Netflix Is Still Missing One Thing: An Identity," *Washington Post*, 22 August 2017, https://www.washingtonpost.com/entertainment/tv/in-its-bid-to-be-everything-netflix-is-still-missing-one-thing-an-identity/2017/08/22/2bb74290-82d9-11e7-b359-15a3617c767b_story.html?utm_term=.0c2486b06300.

8. "Ted Sarandos, Chief Content Officer, Netflix," in *Distribution Revolution: Conversations about the Digital Future of Film and Television*, ed. Michael Curtin, Jennifer Holt, and Kevin Sanson (Oakland: University of California Press, 2014), 144–45 (emphasis in original).

9. "There's Never Enough TV on Netflix," press release, 9 February 2017, https://media.netflix.com/en/press-releases/theres-never-enough-tv-on-netflix.

10. See, for example, Ben Chapman, "How Netflix Took Over the World: Video Streaming Membership Grows to 93.8 Million," *Independent*, 19 January 2017, http://www.independent.co.uk/news/business/news/netflix-video-streaming-93-million-membership-global-6-billion-spending-original-content-series-a7535386.html; Mark Sweney, "Netflix Tops 100m Subscribers as It Draws Worldwide Audience," *Guardian*, 18 July 2017, https://www.theguardian.com/media/2017/jul/18/netflix-tops-100m-subscribers-international-customers-sign-up; John Koblin, "Netflix Says It Will Spend Up to $8 Billion on Content Next Year," *New York Times*, 16 October 2017, https://www.nytimes.com/2017/10/16/business/media/netflix-earnings.html. For more on the rise of discourse on box office see Jon Lewis, "The Perfect Money Machine(s): George Lucas, Steven Spielberg, and Auteurism in the New Hollywood," in *Looking Past the Screen: Case Studies in American Film History and Method* (Durham, NC: Duke University Press, 2007), 61–86.

11. See, for example, Ellen E. Jones, "Send In Brad Pitt: Netflix Gets Out the Big Guns as It Declares War on Cinema," *Guardian*, 21 May 2017, https://www.theguardian.com/film/2017/may/21/war-machine-netflix-home-movie-invasion-brad-pitt-david-michod; Mark Sweney, "Netflix and Amazon 'Will Overtake UK Cinema Box Office Spending by 2020,'" *Guardian*, 14 June 2017, https://www.theguardian.com/media/2017/jun/14/netflix-amazon-uk-cinema-box-office-film-dvd-blu-ray-pwc.

12. See Xavier Amatriain and Justin Basilico, "Netflix Recommendations: Beyond the 5 Stars (Part 1)," *Medium: Netflix Technology Blog*, 5 April 2012, https://medium.com/netflix-techblog/netflix-recommendations-beyond-the-5-stars-part-1-55838468f429; Carlos A. Gomez-Uribe and Neil Hunt, "The Netflix Recommender System: Algorithms, Business Value, and Innovation," *ACM Transactions on Management Information Systems* 6, no. 4, article 13 (2015): 1, 7.

13. "The Story of Netflix, from One of Its Co-Founders," *On Point*, aired 7 November 2019, on NPR. See also Marc Randolph, *That Will Never Work: The Birth of Netflix and the Amazing Life of an Idea* (London: Endeavour, 2019), esp. 4–6.

14. "About Netflix," https://media.netflix.com/en/about-netflix.

15. Neil Monahan and Brandon Griggs, "Why 2.7 Million Americans Still Get Netflix DVDs in the Mail," *CNN*, 4 April 2019, https://edition.cnn.com/2019/04 /04/media/netflix-dvd-subscription-mail-trnd/index.html.

16. See Amatriain and Basilico, "Netflix Recommendations (Part 1)"; Gomez-Uribe and Hunt, "The Netflix Recommender System," 7; Joe Nocera, "Can Netflix Survive in the World It Created?," *New York Times*, 15 June 2016, https:// www.nytimes.com/2016/06/19/magazine/can-netflix-survive-in-the-new-world -it-created.html; Willy Shih and Stephen Kaufman, "Netflix in 2011," *Harvard Business School Case Study*, no. 615-007, 19 August 2014, 13; Pedro Domingos, *The Master Algorithm: How the Quest for the Ultimate Learning Machine Will Remake Our World* (New York: Basic Books, 2015), 184; Joe Weinman, *Digital Disciplines: Attaining Market Leadership via the Cloud, Big Data, Social, Mobile, and the Internet of Things* (Hoboken, NJ: Wiley, 2015), 176, 207, 218.

17. Quote from Lisa Gitelman, *Always Already New: Media, History, and the Data of Culture* (Cambridge, MA: MIT Press, 2006), 10.

18. Shih and Kaufman, "Netflix in 2011," 2.

19. Gina Keating, *Netflixed: The Epic Battle for America's Eyeballs*, 2nd rev. ed. (New York: Portfolio/Penguin, 2013), 21.

20. Keating, *Netflixed*, 24. For more on Maltin's *Movie Guide* and its function in the 1990s, see Daniel Herbert, *Videoland: Movie Culture at the American Video Store* (Berkeley: University of California Press, 2014), 183–217.

21. Keating, *Netflixed*, 23–24, 36.

22. Keating, *Netflixed*, 27.

23. "Ted Sarandos, Chief Content Officer, Netflix," 135.

24. Shih and Kaufman, "Netflix in 2011," 5; Keating, *Netflixed*, 36.

25. Keating, *Netflixed*, 36–37, 61; Shih and Kaufman, "Netflix in 2011," 5–6.

26. Shih and Kaufman, "Netflix in 2011," 5–6.

27. See Michael Bhaskar, "In the Age of the Algorithm, the Human Gatekeeper Is Back," *Guardian*, 30 September 2016, https://www.theguardian.com/technology /2016/sep/30/age-of-algorithm-human-gatekeeper.

28. Shih and Kaufman, "Netflix in 2011," 6, 10; Keating, *Netflixed*, 185. See also Michael D. Smith and Rahul Telang, *Streaming, Sharing, Stealing: Big Data and the Future of Entertainment* (Cambridge, MA: MIT Press, 2016), 128, 130.

29. Computer scientist Pattie Maes, quoted in Clive Thompson, "If You Liked This, You're Sure to Love That," *New York Times*, 21 November 2008, https://www .nytimes.com/2008/11/23/magazine/23Netflix-t.html.

30. Keating, *Netflixed*, 37, 62.

31. Dietmar Jannach et al., *Recommender Systems: An Introduction* (Cambridge: Cambridge University Press, 2011), 161–65.

32. Pierre Bourdieu, *Distinction: A Social Critique of the Judgment of Taste*, trans. Richard Nice (London: Routledge, 1984 [original French-language publication: 1979]).

33. Thompson, "If You Liked This, You're Sure to Love That."

34. Xia Ning, Christian Desrosiers, and George Karypis, "A Comprehensive Survey of Neighborhood-Based Recommendation Methods," in *Recommender Systems Handbook*, 2nd rev. ed., ed. Francesco Ricci, Lior Rokach, and Bracha Shapira (New York: Springer, 2015), 37; Jannach et al., *Recommender Systems*, 20.

35. Laurie J. Flynn, "Like This? You'll Hate That. (Not All Web Recommendations Are Welcome.)" *New York Times*, 23 January 2006, http://www.nytimes.com/2006/01/23/technology/like-this-youll-hate-that-not-all-web-recommendations-are.html.

36. Keating, *Netflixed*, 186.

37. Amatriain and Basilico, "Netflix Recommendations (Part 1)."

38. Smith and Telang, *Streaming, Sharing, Stealing*, 10.

39. Neta Alexander, "Catered to Your Future Self: Netflix's 'Predictive Personalization' and the Mathematization of Taste," in *The Netflix Effect: Technology and Entertainment in the 21st Century*, ed. Kevin McDonald and Daniel Smith-Rowsey (New York: Bloomsbury, 2016), 90. She adopts the terms from Eli Pariser, *The Filter Bubble: What the Internet Is Hiding from You* (London: Penguin, 2011), 117.

40. Casey Johnston, "Netflix Never Used Its $1 Million Algorithm Due to Engineering Costs," *Wired*, 16 April 2012, https://www.wired.com/2012/04/netflix-prize-costs/.

41. Gomez-Uribe and Hunt, "The Netflix Recommender System," 2.

42. Gomez-Uribe and Hunt, "The Netflix Recommender System," 6–7, 9.

43. Amatriain and Basilico, "Netflix Recommendations (Part 1)." The other important algorithmic type is restricted Boltzmann machines (RBMs), which help reduce the complexity of factorization; simply put, they learn users' preferences by calculating the *probabilities* of liking certain latent factors. Machine-learning devices that operate on the model of human-brain neural networks, they help cultural-product recommender systems like Netflix's predict suitable content and reduce the complexity and noise generated by other algorithms. Even after deploying hundreds of algorithms that apply various weights, programmers need to arrange the various weightings by applying meta-weightings to how much and in which cases to calibrate each weighting. For example, RBMs remain more useful when a film or a user has a low number of ratings; matrix factorization methods become more effective when content or a user has a high number of ratings. See Domingos, *The Master Algorithm*, 103–4. See also: Ruslan Salakhutdinov, Andriy Mnih, and Geoffrey Hinton, "Restricted Boltzmann Machines for Collaborative Filtering," in *Proceedings of the 24th International Conference on Machine Learning*, edited by Zoubin Ghahramani (New York: ACM, 2007), 791–98; Edwin Chen, "Introduction to Restricted Boltzmann Machines," *Edwin Chen Blog*, http://blog.echen.me/2011/07/18/introduction-to-restricted-boltzmann-machines/; Edwin Chen,

"Winning the Netflix Prize: A Summary," *Edwin Chen Blog*, http://blog.echen.me /2011/10/24/winning-the-netflix-prize-a-summary/.

44. Domingos, *The Master Algorithm*, 215.

45. See Thompson, "If You Liked This, You're Sure to Love That"; Yehuda Koren and Robert Bell, "Advances in Collaborative Filtering," in *Recommender Systems Handbook*, 2nd rev. ed., ed. Francesco Ricci, Lior Rokach, and Bracha Shapira (New York: Springer, 2015), 77–93; Jannach et al., *Recommender Systems*, 26–31; Chen, "Winning the Netflix Prize." See also Yehuda Koren, Robert Bell, and Chris Volinsky, "Matrix Factorization for Recommender Systems," *Computer* no. 8 (2009): 30–37; Yehuda Koren, "Factorization Meets the Neighborhood: A Multifaceted Collaborative Filtering Model," *Proceedings of the 14th ACM SIGKDD International Conference on Knowledge Discovery and Data Mining* (New York: ACM, 2008), 426–34.

46. Ben Fritz, "Cadre of Film Buffs Helps Netflix Viewers Sort Through the Clutter," *Los Angeles Times*, 3 September 2012, http://articles.latimes.com/2012/sep /03/business/la-fi-0903-ct-netflix-taggers-20120903; see also Alexis C. Madrigal, "How Netflix Reverse Engineered Hollywood," *Atlantic*, 2 January 2014, https:// www.theatlantic.com/technology/archive/2014/01/how-netflix-reverse-engineered -hollywood/282679/.

47. Madrigal, "How Netflix Reverse Engineered Hollywood."

48. Gomez-Uribe and Hunt, "The Netflix Recommender System," 4.

49. A 2012 Netflix blog, quoted in Madrigal, "How Netflix Reverse Engineered Hollywood."

50. See, for example, Nava Tintarev and Judith Masthoff, "Explaining Recommendation: Design and Evaluation," in *Recommender Systems Handbook*, 2nd rev. ed., ed. Francesco Ricci, Lior Rokach, and Bracha Shapira (New York: Springer, 2015), 377; see also Nava Tintarev, "Explanations of Recommendations," *Proceedings of the 2007 ACM Conference on Recommender Systems* (New York: ACM, 2007), 203–6.

51. See, for example, Jannach et al., *Recommender Systems*, 7, 143–47.

52. Jonathan L. Herlocker, Joseph A. Konstan, and John Riedl, "Explaining Collaborative Filtering Recommendations," *Proceedings of the 2000 ACM Conference on Computer Supported Cooperative Work* (New York: ACM, 2000), 241–50; John O'Donovan and Barry Smyth, "Trust in Recommender Systems," *Proceedings of the 10th International Conference on Intelligent User Interfaces* (New York: ACM, 2005), 167–74.

53. Gomez-Uribe and Hunt, "The Netflix Recommender System," 5.

54. See, for example, James F. English, "The Economics of Cultural Awards," in *Handbook of the Economics of Art and Culture*, vol. 2, ed. Victoria A. Ginsburgh and David Throsby (Oxford: North-Holland, 2014), 119–43; Dean Keith Simonton, "Cinematic Success Criteria and Their Predictors: The Art and Business of the Film Industry," *Psychology and Marketing* 26, no. 5 (2009): 400–420.

55. See, for example, Theodor Geiger, "A Radio Taste of Musical Taste," *Public Opinion Quarterly* 14, no. 3 (1950): 453–60.

56. Tom Vanderbilt, *You May Also Like: Taste in an Age of Endless Choice* (London: Simon and Schuster, 2016), 223.

57. See Anthony Jameson et al., "Human Decision Making and Recommender Systems," in *Recommender Systems Handbook*, 2nd rev. ed., ed. Francesco Ricci, Lior Rokach, and Bracha Shapira (New York: Springer, 2015), 627–28.

58. For more on these issues, see Tintarev and Masthoff, "Explaining Recommendations," 358–65.

59. This is Noël Carroll's pithy definition of criticism from his *On Criticism* (New York: Routledge, 2009), 8.

60. Tintarev and Masthoff, "Explaining Recommendations," 366–76.

61. Tintarev and Masthoff, "Explaining Recommendations," 353.

62. These include: linear regression, logistic regression, elastic nets, singular value decomposition, restricted Boltzmann machines, Markov chains, latent Dirichlet allocation, association rules, gradient boosted decision trees, random decision forests, clustering techniques such as simple k-means, affinity propagation, and matrix factorization. See Xavier Amatriain and Justin Basilico, "Netflix Recommendations: Beyond the 5 Stars (Part 2)," *Medium: Netflix Technology Blog*, 19 June 2012, https://medium.com/netflix-techblog/netflix-recommendations-beyond -the-5-stars-part-2-d9b96aa399f5.

63. Jameson et al., "Human Decision Making and Recommender Systems," 637–42; Jannach et al., *Recommender Systems*, 236.

64. Many supervised classification methods can be used for ranking. Typical choices include logistic regression, support vector machines, neural networks, or decision tree–based methods such as gradient boosted decision trees. On the other hand, a great number of algorithms specifically designed for learning to rank have appeared in recent years such as RankSVM or RankBoost. See Amatriain and Basilico, "Netflix Recommendations (Part 2)." See also Gomez-Uribe and Hunt, "The Netflix Recommender System," 5.

65. Amatriain and Basilico, "Netflix Recommendations (Part 1)."

66. Amatriain and Basilico, "Netflix Recommendations (Part 1)."

67. Amatriain and Basilico, "Netflix Recommendations (Part 1)."

68. Gomez-Uribe and Hunt, "The Netflix Recommender System," 2–3.

69. Gomez-Uribe and Hunt, "The Netflix Recommender System," 3–4.

70. Amatriain and Basilico, "Netflix Recommendations (Part 1)."

71. Gomez-Uribe and Hunt, "The Netflix Recommender System," 3.

72. For more on these bugs of human pattern-detection and causal logic, see Weinman, *Digital Disciplines*, 189.

73. Domingos, *The Master Algorithm*, 71–72.

74. For critiques of overfitting, see especially Pariser, *The Filter Bubble*; Joseph Turow, *The Daily You: How the New Advertising Industry Is Defining Your Identity and Your Worth* (New Haven, CT: Yale University Press, 2011); Cass R. Sunstein, *Republic.com 2.0* (Princeton, NJ: Princeton University Press, 2007). For computer science discussions and remedies of the problem, see Jannach et al., *Recommender Systems*, 76; Ning, Desrosiers, and Karypis, "A Comprehensive Survey of

Neighborhood-Based Recommendation Methods," 39; and Pablo Castells, Neil J. Hurley, and Saul Vargas, "Novelty and Diversity in Recommender Systems," in *Recommender Systems Handbook*, 2nd rev. ed., ed. Francesco Ricci, Lior Rokach, and Bracha Shapira (New York: Springer, 2015), 881–918.

75. Domingos, *The Master Algorithm*, 76.

76. Amatriain and Basilico, "Netflix Recommendations (Part 2)."

77. Amatriain and Basilico, "Netflix Recommendations (Part 1)."

78. See Vanderbilt, *You May Also Like*, 28–29; Andrea Maier, Zata Vickers, and J. Jeffrey Inman, "Sensory-Specific Satiety, Its Crossovers, and Subsequent Choice of Potato Chip Flavors," *Appetite* 49, no. 2 (2007): 419–28.

79. See, for example, Richard Corliss, "All Thumbs or, Is There a Future for Film Criticism?," *Film Comment* 26, no. 2 (1990): 14–18.

80. Tom Vanderbilt, "Now Netflix Is All Thumbs," *New York Times*, 31 March 2017, https://www.nytimes.com/2017/03/31/opinion/now-netflix-is-all-thumbs .html; Vanderbilt, *You May Also Like*.

81. Chen, "Winning the Netflix Prize."

82. Vanderbilt, *You May Also Like*, 57.

83. Vanderbilt, *You May Also Like*, 58.

84. See Vanderbilt, *You May Also Like*, 54.

85. Francesco Ricci, Lior Rokach, and Bracha Shapira, "Recommender Systems: Introduction and Challenges," in *Recommender Systems Handbook*, 2nd rev. ed., ed. Francesco Ricci, Lior Rokach, and Bracha Shapira (New York: Springer, 2015), 1.

86. See Herbert A. Simon, "A Behavioral Model of Rational Choice," *Quarterly Journal of Economics* 69, no. 1 (1955): 99–118; Gerd Gigerenzer and Reinhard Selten, eds., *Bounded Rationality: The Adaptive Toolbox* (Cambridge, MA: MIT Press, 2001).

87. Gomez-Uribe and Hunt, "The Netflix Recommender System," 7.

88. See Barry Schwartz, *The Paradox of Choice: Why More Is Less* (New York: Ecco, 2004); Richard H. Thaler and Cass B. Sunstein, *Nudge: Improving Decisions about Health, Wealth, and Happiness* (New Haven, CT: Yale University Press, 2008).

89. For a brief discussion of Netflix's queue as a tool of surplus management, see Amanda D. Lotz, *The Television Will Be Revolutionized*, 2nd rev. ed. (New York: New York University Press, 2014), 74.

90. Richard Peterson and Roger Kern, "Changing Highbrow Taste: From Snob to Omnivore," *American Sociological Review* 61, no. 5 (1996): 900–907.

91. Ramon Lobato, *Netflix Nations: The Geography of Digital Distribution* (New York: New York University Press, 2019), 20–21, 14–42 (quote on 42).

CHAPTER 4. UNPACKING NETFLIX'S MYTH OF BIG DATA

1. Joe Weinman, *Digital Disciplines: Attaining Market Leadership via the Cloud, Big Data, Social, Mobile, and the Internet of Things* (Hoboken, NJ: Wiley, 2015), 181.

2. Sarah Arnold, "Netflix and the Myth of Choice/Participation/Autonomy," in *The Netflix Effect: Technology and Entertainment in the 21st Century*, ed. Kevin McDonald and Daniel Smith-Rowsey (New York: Bloomsbury, 2016), 50. For another argument in this vein, see Annette Markham, Simona Stavrova, and Max Schlüter, "Netflix, Imagined Affordances, and the Illusion of Control," *Netflix at the Nexus: Content, Practice, and Production in the Age of Streaming Television*, ed. Amber M. Buck and Theo Plothe (Bern: Peter Lang, 2019), 29–46.

3. Arnold, "Netflix and the Myth of Choice/Participation/Autonomy," 55.

4. See John Cheney-Lippold, "A New Algorithmic Identity: Soft Biopolitics and the Modulation of Control," *Theory, Culture and Society* 28, no. 6 (2011): 164–81; John Cheney-Lippold, *We Are Data: Algorithms and the Making of Our Digital Selves* (New York: New York University Press, 2017); Greg Elmer, *Profiling Machines: Mapping the Personal Information Economy* (Cambridge, MA: MIT Press, 2004); Blake Hallinan and Ted Striphas, "Recommended for You: The Netflix Prize and the Production of Algorithmic Culture," *New Media and Society* 18, no. 1 (2014): 117–37.

5. Neta Alexander, "Catered to Your Future Self: Netflix's 'Predictive Personalization' and the Mathematization of Taste," in *The Netflix Effect: Technology and Entertainment in the 21st Century*, ed. Kevin McDonald and Daniel Smith-Rowsey (New York: Bloomsbury, 2016), 81–82.

6. Alexander, "Catered to Your Future Self," 84.

7. Alexander, "Catered to Your Future Self," 88.

8. Alexander, "Catered to Your Future Self," 88, 91.

9. Daniel Smith-Rowsey, "Imaginative Indices and Deceptive Domains: How Netflix's Categories and Genres Redefine the Long Tail," in *The Netflix Effect: Technology and Entertainment in the 21st Century*, ed. Kevin McDonald and Daniel Smith-Rowsey (New York: Bloomsbury, 2016), 64, 67. See also Jessica Balanzategui, Liam Burke, and Dan Golding, "Recommending a New System: An Audience-Based Approach to Film Categorisation in the Digital Age," *Participations* 15, no. 2 (2018): 297–328.

10. Ted Striphas, "Algorithmic Culture," *European Journal of Cultural Studies* 18, nos. 4–5 (2015): 395.

11. Striphas, "Algorithmic Culture," 405, 408. Striphas's references to "sociotechnical assemblages" and the automation of cultural decision-making processes pertain to the thoughts of, respectively, Tarleton Gillespie and Vilém Flusser.

12. Hallinan and Striphas, "Recommended for You," 122.

13. Hallinan and Striphas, "Recommended for You," 131.

14. Hallinan and Striphas, "Recommended for You," 117, 120.

15. Hallinan and Striphas, "Recommended for You," 131.

16. Striphas, "Algorithmic Culture," 397.

17. Striphas, "Algorithmic Culture," 407.

18. danah boyd and Kate Crawford, "Critical Questions for Big Data: Provocations for a Cultural, Technological, and Scholarly Phenomenon," *Information, Communication and Society* 15, no. 5 (2012): 663.

19. Tarleton Gillespie, "The Relevance of Algorithms," in *Media Technologies: Essays on Communication, Materiality, and Society*, ed. Tarleton Gillespie, Pablo J. Boczkowski, and Kirsten A. Foot (Cambridge, MA: MIT Press, 2014), 180. Gillespie borrows the term "performed backstage" from Stephen Hilgartner, *Science on Stage: Expert Advice as Public Drama* (Stanford, CA: Stanford University Press, 2000).

20. Gina Keating, *Netflixed: The Epic Battle for America's Eyeballs*, 2nd rev. ed. (New York: Portfolio/Penguin, 2013), 186–87.

21. Katie Hafner, "And If You Liked the Movie, a Netflix Contest May Reward You Handsomely," *New York Times*, 2 October 2016, http://www.nytimes.com /2006/10/02/technology/02netflix.html.

22. Keating, *Netflixed*, 196.

23. To be sure, Netflix has provided a number of reasons for not implementing the winning model. Beyond a cost-benefit analysis that the resulting engineering and data processing would have been prohibitively expensive, the company has cited the pending transition to streaming and the concomitant increased ability to collect behavioral data as a further rationale. Improving star-rating predictions, the putative objective of the competition, lost importance under the new model. See Xavier Amatriain and Justin Basilico, "Netflix Recommendations: Beyond the 5 Stars (Part 1)," *Medium: Netflix Technology Blog*, 5 April 2012, https:// medium.com/netflix-techblog/netflix-recommendations-beyond-the-5-stars-part-1 -55838468f429.

24. Keating, *Netflixed*, 196–97, 184. See also Ken Auletta's account, which, among many others, largely echoes these observations: "Outside the Box: Netflix and the Future of Television," *New Yorker*, 3 February 2014, https://www.newyorker .com/magazine/2014/02/03/outside-the-box-2.

25. Auletta, "Outside the Box."

26. Weinman, *Digital Disciplines*, 197–98, 200–201.

27. Michael Gubbins, "Digital Revolution: Active Audiences and Fragmented Consumption," in *Digital Disruption: Cinema Moves On-line*, ed. Dina Iordanova and Stuart Cunningham (St. Andrews, UK: St. Andrews Film Studies, 2012), 67–68.

28. Gubbins, "Digital Revolution," 76.

29. Joe Nocera, "Can Netflix Survive in the World It Created?," *New York Times*, 15 June 2016, https://www.nytimes.com/2016/06/19/magazine/can-netflix -survive-in-the-new-world-it-created.html; Auletta, "Outside the Box."

30. Mario Cibelli, Marathon Partners, quoted in Keating, *Netflixed*, 136.

31. "Letter to Shareholders (Q3, 2012)," 23 October 2012, https://www.sec.gov /Archives/edgar/data/1065280/000106528012000015/nflx-093012xex991.htm.

32. "Ted Sarandos, Chief Content Officer, Netflix," in *Distribution Revolution: Conversations about the Digital Future of Film and Television*, ed. Michael Curtin, Jennifer Holt, and Kevin Sanson (Oakland: University of California Press, 2014), 138.

33. Macmillan CEO John Sargent, paraphrased in Michael D. Smith and Rahul Telang, *Streaming, Sharing, Stealing: Big Data and the Future of Entertainment* (Cambridge, MA: MIT Press, 2016), 147–48.

34. For more on engineering culture, see Nathan Ensmenger, *The Computer Boys Take Over: Computers, Programmers, and the Politics of Technical Expertise* (Cambridge, MA: MIT Press, 2010).

35. Michael Wolff, *Television Is the New Television: The Unexpected Triumph of Old Media in the Digital Age* (New York: Portfolio/Penguin, 2015), esp. 4–5.

36. Wolff, *Television Is the New Television*, 10–11.

37. Wolff, *Television Is the New Television*, 91.

38. See Erik Adams, "Netflix Says Nielsen's Data Is Inaccurate—So Why Doesn't It Release the Numbers Its Damn Self?," *AV Club*, 18 October 2017, https://www.avclub.com/netflix-says-nielsen-s-data-is-inaccurate-so-why-doesn-1819652138 (quote from this source); Jacob Oller, "Nielsen Begins Tracking Netflix Viewership Data," *Forbes*, 18 October 2017, https://www.forbes.com/sites/jacoboller/2017/10/18/nielsen-begins-tracking-netflix-viewership-data/#33d8c64725eb; Rani Molla, "TV Producers Can Now See How Many People Are Watching Netflix Shows," *Recode*, 18 October 2017, https://www.recode.net/2017/10/18/16495130/netflix-the-defenders-6-1-million-viewers-season-premiere.

39. Katie Urban, "Netflix Study Reveals If You Want to Feel Closer to Your Teen, Watch Their Shows," *Netflix Media Center*, 25 April 2017, https://media.netflix.com/en/press-releases/netflix-study-reveals-if-you-want-to-feel-closer-to-your-teen-watch-their-shows; Katie Urban, "Netflix Cheating Is on the Rise Globally and Shows No Signs of Stopping," *Netflix Media Center*, 13 February 2017, https://media.netflix.com/en/press-releases/netflix-cheating-is-on-the-rise-globally-and-shows-no-signs-of-stopping; "Netflix Boldly Goes Where No Man Has Gone Before, Revealing *Star Trek* Fans' Favorite Episodes," *Netflix Media Center*, 8 September 2017, https://media.netflix.com/en/press-releases/netflix-boldly-goes-where-no-man-has-gone-before-revealing-star-trek-fans-favorite-episodes; Erin Dwyer, "Decoding the Defenders: Netflix Unveils the Gateway Shows That Lead to a Heroic Binge," *Netflix Media Center*, 22 August 2017, https://media.netflix.com/en/press-releases/decoding-the-defenders-netflix-unveils-the-gateway-shows-that-lead-to-a-heroic-binge; Myles Worthington, "Sneaking Is the New Bingeing: Moms Get More Creative Than Ever by Squeezing in Time to Stream Netflix," *Netflix Media Center*, 11 July 2017, https://media.netflix.com/en/press-releases/sneaking-is-the-new-bingeing-moms-get-more-creative-than-ever-by-squeezing-in-time-to-stream-netflix; "How Netflix Hack Helps Parents Blow Out the Candles on the Birthday Party Hijinks," *Netflix Media Center*, 14 September 2017, https://media.netflix.com/en/press-releases/new-netflix-hack-helps-parents-blow-out-the-candles-on-the-birthday-party-hijinks. The parenting-related press releases indeed appeared recycled on lifestyle blogs like *Ma Belle Vie* and *The Mommy Mix*. For more on Netflix's early-user demographic profile, see Keating, *Netflixed*, 37. Quote about the Stream Team derives from: "Netflix Stream Team: The Beginning," *Dad You Geek*, 5 May 2015, https://www.dadyougeek.co.uk/new-blog/2015/4/5/netflix-stream-team-the-beginning.

40. "Joining the Netflix Stream Team," *What Katy Said: About Motherhood and Life*, 7 June 2016, https://www.whatkatysaid.com/joining-netflix-stream-team/.

41. Niraj Chokshi, "Americans Are Watching Netflix at Work and in the Bathroom," *New York Times*, 17 November 2017, https://www.nytimes.com/2017/11/17/business/media/watch-netflix-at-work.html. The press release basis for this story is: Sean Flynn, "When Bingeing Goes Public, Private Behaviors Are Exposed and Social Norms Are Shelved," *Netflix Media Center*, 14 November 2017, https://media.netflix.com/en/press-releases/when-bingeing-goes-public-private-behaviors-are-exposed-and-social-norms-are-shelved.

42. David Carr, "Giving Viewers What They Want," *New York Times*, 24 February 2013, http://www.nytimes.com/2013/02/25/business/media/for-house-of-cards-using-big-data-to-guarantee-its-popularity.html.

43. See, for example, Auletta, "Outside the Box"; Nocera, "Can Netflix Survive in the World It Created?"; Weinman, *Digital Disciplines*, 201; James G. Webster, *The Marketplace of Attention: How Audiences Take Shape in a Digital Age* (Cambridge, MA: MIT Press, 2014), 60; Hallinan and Striphas, "Recommended for You," 128–29.

44. Jane Martinson, "Netflix's Ted Sarandos: 'We Like Giving Great Storytellers Big Canvases,'" *Guardian*, 15 March 2015, https://www.theguardian.com/media/2015/mar/15/netflix-ted-sarandos-house-of-cards.

45. See Smith and Telang, *Streaming, Sharing, Stealing*, 149.

46. See Anita Elberse, "MRC's House of Cards," *Harvard Business School Case Study*, no. 515-003, 16 January 2015.

47. Clive Thompson, "If You Liked This, You're Sure to Love That," *New York Times*, 21 November 2008, https://www.nytimes.com/2008/11/23/magazine/23Netflix-t.html.

48. Keating, *Netflixed*, 192.

49. Nocera, "Can Netflix Survive in the World It Created?"

50. Alexis C. Madrigal, "How Netflix Reverse Engineered Hollywood," *Atlantic*, 2 January 2014, https://www.theatlantic.com/technology/archive/2014/01/how-netflix-reverse-engineered-hollywood/282679/.

51. Adam Sherwin, "Netflix and Kill: Bosses 'Ruthless' with Weak Series," *i*, 8 July 2017, 21.

52. Scott Mendelson, "Brad Pitt's Disappointing 'War Machine' Hurts Netflix, Helps Movie Theaters," *Forbes*, 31 May 2017, https://www.forbes.com/sites/scottmendelson/2017/05/31/brad-pitts-disappointing-war-machine-hurts-netflix-helps-movie-theaters/#7bae8c4633c2.

53. Furthermore, taste awareness helps Netflix better manage the acquisition (and commissioning) of films and series by knowing how many titles of which sort the company needs to acquire in order to maintain customer satisfaction and, theoretically at least, to better gauge and ultimately reduce how much Netflix is willing to pay in licensing fee negotiations. Netflix claims to use its data to leverage acquisition costs. According to Sarandos: "We use it to indicate how much I should or shouldn't pay. In other words, if I can get an enormous amount of viewing, I'll pay an enormous amount of money" ("Ted Sarandos, Chief Content Officer," 136). Nevertheless, there are reasons to doubt such pronouncements: industry insiders maintain that the company vastly overpays for content.

54. First passage quoted in Victor Luckerson, "2015 Will Be the Year Netflix Goes 'Full HBO,'" *Time*, 20 January 2015, http://time.com/collection-post/3675669 /netflix-hbo/. On this issue, see also Nocera, "Can Netflix Survive in the World It Created"; "Ted Sarandos, Chief Content Officer," 140. The second quotation derives from "Ted Sarandos, Chief Content Officer," 141. A number of commentators have noted the shift in Netflix's business model, including Auletta, "Outside the Box"; Keating, *Netflixed*; Weinman, *Digital Disciplines*; Wolff, *Television Is the New Television*.

55. See Paul Bond and Tim Appelo, "How the Assault on Netflix Will Shake Out," *Hollywood Reporter*, 25 March 2011, 36–39.

56. "The Television Will Be Revolutionised," *Economist*, 30 June 2018, 17–20.

57. See Brooks Barnes, "How Disney Wants to Take on Netflix with Its Own Streaming Service," *New York Times*, 8 August 2017, https://www.nytimes.com /2017/08/08/business/media/disney-streaming-service.html; Brooks Barnes, "Disney Makes Deal for 21st Century Fox, Reshaping Entertainment Landscape," *New York Times*, 14 December 2017, https://www.nytimes.com/2017/12/14/business /dealbook/disney-fox-deal.html; John Koblin, "Apple's Big Spending Plan to Challenge Netflix Takes Shape," *New York Times*, 17 March 2019, https://www.nytimes .com/2019/03/17/business/media/apple-content-hollywood.html; Brooks Barnes, "Disney Moves from Behemoth to Colossus with Closing of Fox Deal," *New York Times*, 20 March 2019, https://www.nytimes.com/2019/03/20/business/media/walt -disney-21st-century-fox-deal.html; Matthew Ball, "Streaming Video Will Soon Look Like the Bad Old Days of TV," *New York Times*, 22 August 2019, https:// www.nytimes.com/2019/08/22/opinion/netflix-hulu-cable.html.

58. See, for example, Bond and Appelo, "How the Assault on Netflix Will Shake Out"; Stuart Cunningham and Jon Silver, *Screen Distribution and the New King Kongs of the Online World* (Basingstoke, UK: Palgrave Macmillan, 2013).

59. Edmund Lee, "Everyone You Know Just Signed up for Netflix," *New York Times*, 21 April 2020, https://nyti.ms/3cDh1cW; Mark Sweney, "Disney Forecast to Steal Netflix's Crown as Biggest Streaming Firm," *Guardian*, 14 March 2021, https://www.theguardian.com/film/2021/mar/14/disney-forecast-to-steal-netflix -crown-as-worlds-biggest-streaming-firm.

60. On this point, see Nocera, "Can Netflix Survive in the World It Created?"

61. Simon Usborne, "Netflix's 'New World Order': A Streaming Giant on the Brink of Global Domination," *Guardian*, 17 April 2018, https://www.theguardian .com/media/2018/apr/17/netflixs-new-world-order-a-streaming-giant-on-the-brink -of-global-domination. On Netflix's precarious finances, see Auletta, "Outside the Box"; "The Television Will Be Revolutionised," 17–20; Nocera, "Can Netflix Survive in the World It Created?"; Mark Sweney, "Netflix Has Revolutionised Television. But Is Its Crown Starting to Slip?," *Guardian*, 21 July 2018, https://www .theguardian.com/media/2018/jul/21/netflix-crown-beginning-to-slip-subscriber -numbers; Mark Sweney, "Amazon Prime Video's Growth Outpaces Netflix in UK," *Guardian*, 3 May 2018, https://www.theguardian.com/media/2018/may/03/amazon -prime-video-outpaces-netflix-in-uk-on-demand-market.

62. See, for example, Brian Fung, "Why Netflix Is Raising Prices Again and What It Could Mean for Cord-Cutting," *Washington Post*, 5 October 2017, https://www .washingtonpost.com/news/the-switch/wp/2017/10/05/why-netflix-is-raising-prices -again-and-what-it-could-mean-for-cord-cutting/?utm_term=.f8d9449dda97; Andrew Griffin, "Netflix Price Increase: Subscription Fee Automatically Rises by 20%," *Independent*, 30 May 2019, https://www.independent.co.uk/life-style/gadgets -and-tech/news/netflix-price-increase-fee-cost-how-much-rise-stream-a8936546 .html; Mark Sweney, "Netflix Subscriber Slowdown Could Mark Streaming Giant's Peak," *Guardian*, 17 July 2018, https://www.theguardian.com/media/2018/jul/17 /netflix-subscriber-slowdown-could-mark-streaming-giants-peak; Steven Zeitchik, "Netflix Stock Plummets After Streaming Video Company Misses Subscriber Target," *Washington Post*, 17 July 2018, https://www.washingtonpost.com/business/2018 /07/16/netflix-subscriber-growth-slows-panicking-wall-street/?noredirect=on.

63. For a discussion of Netflix's Qwikster debacle and other poor customer service decisions, see Keating, *Netflixed*, 120, 177, 247–54; for an exemplary diagnosis of Netflix's "creepy" data-collection methods, see Hayley Tsukuyama, "Netflix's 'Creepy' Tweet Reminds Us All How Closely It's Watching Us," *Washington Post*, 11 December 2017, https://www.washingtonpost.com/news/the-switch/wp/2017/12 /11/netflixs-creepy-tweet-reminds-us-all-how-closely-its-watching-us/?utm_term= .70994c35ce1a; Sapna Maheshwari, "Netflix and Spotify Ask: Can Data Mining Make for Cute Ads?," *New York Times*, 17 December 2017, https://www.nytimes .com/2017/12/17/business/media/netflix-spotify-marketing.html?_r=0.

64. See, for instance, Safiya Umoja Noble, *Algorithms of Oppression: How Search Engines Reinforce Racism* (New York: New York University Press, 2018); Cathy O'Neil, *Weapons of Math Destruction: How Big Data Increases Inequality and Threatens Democracy* (New York: Penguin Random House, 2016).

65. Pedro Domingos, *The Master Algorithm: How the Quest for the Ultimate Learning Machine Will Remake Our World* (New York: Basic Books, 2015), 1–2.

66. Arnold, "Netflix and the Myth of Choice/Participation/Autonomy," 57.

67. Gillespie, "The Relevance of Algorithms," 175.

68. Tim Wu, "Netflix's Secret Special Algorithm Is a Human," *New Yorker*, 27 January 2015, https://www.newyorker.com/business/currency/hollywoods-big -data-big-deal; cf. Willy Shih and Stephen Kaufman, "Netflix in 2011," *Harvard Business School Case Study*, no. 615-007, 19 August 2014, 9.

69. Martinson, "Netflix's Ted Sarandos."

70. Quoted in European Audiovisual Observatory, *The Development of the European Market for On-Demand Audiovisual Services* (Strasbourg: European Audiovisual Observatory, 2015), 143.

71. For examples of the widespread consensus about this "problem" both among Netflix critics and media studies more generally, see Hallinan and Striphas, "Recommended for You," e.g., 118; Frank Pasquale, *The Black Box Society: The Secret Algorithms That Control Money and Information* (Cambridge, MA: Harvard University Press, 2015); John Cheney-Lippold, *We Are Data: Algorithms and the Making of Our Digital Selves* (New York: New York University Press, 2017), 4 and

passim; Paul Dourish, "Algorithms and Their Others: Algorithmic Culture in Context," *Big Data and Society* 3, no. 2 (2016): 1–11, esp. 6ff.; Johannes Paßmann and Asher Boersma, "Unknowing Algorithms: On Transparency of Unopenable Black Boxes," in *The Datafied Society: Studying Culture through Data*, ed. Mirko Tobias Schäfer and Karin van Es (Amsterdam: Amsterdam University Press, 2017), 139–46.

72. The first quotations stem from David Wright, *Understanding Cultural Taste: Sensation, Skill and Sensibility* (Basingstoke, UK: Palgrave Macmillan, 2015), 161; the last is the diagnosis of Gillespie, "The Relevance of Algorithms," 176.

73. David Beer, *Popular Culture and New Media: The Politics of Circulation* (Basingstoke, UK: Palgrave Macmillan, 2013), 95.

74. Stuart Ewen, *Captains of Consciousness: Advertising and the Social Roots of the Consumer Culture*, 2nd rev. ed. (New York: Basic Books, 2001), 105–7.

75. See Matthew Freeman, *Industrial Approaches to Media: A Methodological Gateway to Industry Studies* (London: Palgrave Macmillan, 2016), e.g., 112–14.

76. Evgeny Morozov, "Opposing the Exceptionalism of the Algorithm," in *The Datafied Society: Studying Culture through Data*, ed. Mirko Tobias Schäfer and Karin van Es (Amsterdam: Amsterdam University Press, 2017), 246.

77. For the unreliability of entertainment industry data and accounting, even before the digital age, see Sergio Sparviero, "Hollywood Creative Accounting: The Success Rate of Major Motion Pictures," *Media Industries* 2, no. 1 (2015): 19–36.

78. Bernhard Rieder, *Engines of Order: A Mechanology of Algorithmic Techniques* (Amsterdam: Amsterdam University Press, 2020), 9.

79. Grant Blank, *Critics, Ratings, and Society: The Sociology of Reviews* (Lanham, MD: Rowman & Littlefield, 2007), 133.

80. Usborne, "Netflix's 'New World Order.'"

81. These are the most frequently mentioned criticisms of Netflix in consumer surveys; see also Lizzie O'Shea, "What Kind of a Person Does Netflix Favourites Think I Am?," *Guardian*, 16 April 2008, https://www.theguardian.com /commentisfree/2018/apr/16/netflix-favourites-algorithm-recommendations; Jiji Lee, "'What Should I Watch on Netflix?': A New Original Series," *New York Times*, 11 November 2017, https://www.nytimes.com/2017/11/11/opinion/sunday/what -should-i-watch-on-netflix.html.

82. These admitted areas for improvement have included developing better recommendation provision for new members without a long behavioral track record on Netflix; measuring engagement (counting hours favors series, counting discrete numbers of titles favors feature films); and recommending for children, whose tastes change quickly as they grow up. See Carlos A. Gomez-Uribe and Neil Hunt, "The Netflix Recommender System: Algorithms, Business Value, and Innovation," *ACM Transactions on Management Information Systems* 6, no. 4, article 13 (2015): 14–15.

83. Cees van Koppen, Netflix manager public policy, EMEA, "Commercial Strategies in the UK TV Market—Audiences, Technology and Discoverability," presentation at the conference "Competition in the UK TV Market: Consumer Trends, Commercial Strategies and Policy Options," London, 24 October 2017, 21.

84. Amatriain and Basilico, "Netflix Recommendations (Part 1)."

1. Alison N. Novak's small-scale ($n = 27$) examination of millennial university students' feelings about Netflix personalization, conducted by journal analysis, yields important insights regarding this specific age cohort's often blasé attitudes toward recommender systems and personalized recommendations. See Alison N. Novak, "Narrowcasting, Millennials and the Personalization of Genre in Digital Media," in *The Age of Netflix: Critical Essays on Streaming Media, Digital Delivery and Instant Access*, ed. Cory Barker and Myc Wiatrowski (Jefferson, NC: McFarland, 2017), 162–81. Emil Steiner and Kun Xu have made inroads into understanding audiences' motivations for binge-watching with qualitative research that includes: Emil Steiner, "Binge-Watching in Practice: The Rituals, Motives and Feelings of Streaming Video Viewers," in *The Age of Netflix: Critical Essays on Streaming Media, Digital Delivery and Instant Access*, ed. Cory Barker and Myc Wiatrowski (Jefferson, NC: McFarland, 2017), 141–61; and Emil Steiner and Kun Xu, "Binge-Watching Motivates Change: Uses and Gratifications of Streaming Video Viewers Challenge Traditional TV Research," *Convergence* 26, no. 1 (2020): 82–101. An EU-wide large-scale study, *A Profile of Current and Future Audiovisual Audience*, is extremely valuable for its comprehensive treatment of the ways by which European audiences learn about and choose films. Nevertheless, the fieldwork was undertaken in March and April 2013, before the widespread uptake of Netflix across the Continent. Therefore, its results must be read with some caution. See European Commission, *A Profile of Current and Future Audiovisual Audience* (Luxembourg: Publications Office of the European Union, 2014). I consciously chose to replicate some of the EU study's questions in my surveys so as to chart changes in attitudes over time. See the afterword.

2. Benjamin Toff and Rasmus Kleis Nielsen, "'I Just Google It': Folk Theories of Distributed Discovery," *Journal of Communication* 68, no. 3 (2018): 637.

3. The figures in these three sentences refer to the UK results. The US respondents rank the recommendation sources differently, but word of mouth similarly emerges far above critics, and critics in turn rate as significantly more influential than recommender systems.

4. See (in the appendix) the differences between the responses to, on the one hand, survey questions 1 and 2 (which pertain to all means of consumption including cinema, television, DVD, and VOD), and on the other, question 4, which asks the subsample of VOD users only about their VOD use.

5. Participant 6 was revealing in this aspect. He noted that his use of the South Asian VOD service Einthusan included consulting the user ratings that are built into the platform but having to go to YouTube to watch trailers of the films on offer.

6. Other participants, when asked about their preferences, explicitly differentiated between situations in which they had—in the words of Participant 2—"something in mind" and those in which they were "shopping around"—that is, searching based on a tip or browsing. Other participants distinguished, in the same way, between

selecting for an intensive viewing experience and those situations (cooking, eating, about to sleep) in which they expected the audiovisual content to function essentially as white noise. Participants 6 and 34 differentiated between films with which they are familiar (i.e., the new *Star Wars* or Marvel film) and those with which they are unfamiliar and feel more need to research in order to gauge interest.

7. On the "awareness effects" (vs. "persuasive effects") of criticism, see Wen-jing Duan, Bin Gu, and Andrew Whinston, "Do Online Reviews Matter?—An Empirical Investigation of Panel Data," *Decision Support Systems* 45, no. 4 (2008): 1007–16; Greta Hsu and Joel M. Podolny, "Critiquing the Critics: An Approach for the Comparative Evaluation of Critical Schemas," *Social Science Research* 34, no. 1 (2005): 189–214. On news media's "signal function," a related phenomenon, see Walter Lippmann, *Public Opinion* (New York: Macmillan, 1922).

8. See, for example, John Cheney-Lippold, *We Are Data: Algorithms and the Making of Our Digital Selves* (New York: New York University Press, 2017).

9. Viktoria Grzymek and Michael Puntschuh, *Was Europa über Algorithmen weiß und denkt: Ergebnisse einer repräsentativen Bevölkerungsumfrage* (Gütersloh: Bertelsmann Stiftung, 2019). See also Sarah Fischer and Thomas Petersen, *Was Deutschland über Algorithmen weiß und denkt: Ergebnisse einer repräsentativen Bevölkerungsumfrage* (Gütersloh: Bertelsmann Stiftung, 2018), esp. 8, 13–15, 25–27.

10. Aaron Smith, *Public Attitudes toward Computer Algorithms* (Washington, DC: Pew Research Center, 2018).

11. Fischer and Petersen, *Was Deutschland über Algorithmen weiß und denkt*, 8.

12. Fischer and Petersen, *Was Deutschland über Algorithmen weiß und denkt*, esp. 8, 13–15, 25–27.

13. See, for example, BuzzFeed's "Netflix in Real Life," a video clip parody whereby a video shop clerk awkwardly simulates Netflix's recommender system. The spoof lampoons the system's low use-value, repetitive or inaccurate suggestions, and sparse content catalog. See: https://www.youtube.com/watch?v=EglLfaECsdU. For more on Netflix memes, including a brief discussion of memes that suggest Netflix recommendations are "intentionally boring," see Elena Pilipets, "From Netflix Streaming to Netflix and Chill: The (Dis)connected Body of Serial Binge-Viewer," *Social Media and Society* 5, no. 4 (2019).

14. The "don't-care" faction resounded with Novak's findings of millennials' blasé attitudes toward Netflix's personalized recommendations and the underlying technologies. See Novak, "Narrowcasting, Millennials and the Personalization of Genre in Digital Media," 162–81.

15. Toff and Nielsen, "I Just Google It," 654.

16. Novak, "Narrowcasting, Millennials and the Personalization of Genre in Digital Media," 162–81.

17. See Daniel J. O'Keefe, *Persuasion: Theory and Research*, 3rd rev. ed. (Los Angeles: Sage, 2016); Kyung-Hyan Yoo, Ulrike Gretzel, and Markus Zanker, "Source Factors in Recommender System Credibility Evaluation," in *Recommender Systems Handbook*, 2nd rev. ed., ed. Francesco Ricci, Lior Rokach, and Bracha Shapira (New York: Springer, 2015), 689–714; Nava Tintarev and Judith Masthoff,

"Explaining Recommendation: Design and Evaluation," in *Recommender Systems Handbook*, 2nd rev. ed., ed. Francesco Ricci, Lior Rokach, and Bracha Shapira (New York: Springer, 2015), 353–82.

18. Nevertheless, her acceptance of the recommendations is neither one-dimensional nor absolute. When asked whether the match score influenced her decision, she said that it played no significant role: "I don't choose it because of that. I go directly to the list, and as I was telling you, I think I am very visual. So, if the image goes directly to an actor that I know, for example, or the title is very catchy, or if the image looks very cool, then I will go directly and read the synopsis of the movie. If that's something that I really want to watch at that time then I will go, 'Oh, yes, it's fine.' Or, 'Oh, I will add it to my list for me to watch it later,' and then I go for a different option. I select the movies based on the visuals of the poster, of the image on Netflix, [I would say], instead of the matching score."

19. Quoted in "How to Devise the Perfect Recommendation Algorithm," *Economist*, 9 February 2017, https://www.economist.com/special-report/2017/02/09/how-to-devise-the-perfect-recommendation-algorithm.

20. Lizzie O'Shea, "What Kind of a Person Does Netflix Favourites Think I Am?," *Guardian*, 16 April 2018, https://www.theguardian.com/commentisfree/2018/apr/16/netflix-favourites-algorithm-recommendations.

21. See, for example, Fischer and Petersen, *Was Deutschland über Algorithmen weiß und denkt*; Novak, "Narrowcasting, Millennials and the Personalization of Genre in Digital Media," esp. 178.

22. Jonathan Cohn, *The Burden of Choice: Recommendations, Subversion, and Algorithmic Culture* (New Brunswick, NJ: Rutgers University Press, 2019), esp. 94–105.

23. See, for an example of that discourse, Cara Buckley, "When 'Captain Marvel' Became a Target, the Rules Changed," *New York Times*, 13 March 2019, https://www.nytimes.com/2019/03/13/movies/captain-marvel-brie-larson-rotten-tomatoes.html; for a stark empirical indication of the lack of female (and racial minority) representation among film critics in the United States, see Marc Choueiti, Stacy L. Smith, and Katherine Pieper, *Critic's Choice 2: Gender and Race/Ethnicity of Film Reviewers Across 300 Top Films from 2015–2017* (Los Angeles: USC Annenberg Inclusion Initiative, 2018).

24. For more on multitasking and second-screening, see Claire M. Segijn and Anastasia Kononova, "Audience, Media and Cultural Factors as Predictors of Multiscreen Use: A Comparative Study of the Netherlands and the United States," *International Journal of Communication* no. 12 (2018): 4708–30; Jennifer Holt and Kevin Sanson, eds., *Connected Viewing: Selling, Streaming and Sharing Media in the Digital Era* (New York: Routledge, 2014).

25. See, for example, European Commission, *A Profile of Current and Future Audiovisual Audience*, 18–19, which shows how age, gender, and level of activity are less predictive of heavy consumption of films (across all media) than location (i.e., urban) and household income (low-income Europeans consume considerably more films).

26. See Vinod Krishnan et al., "Who Predicts Better?—Results from an Online Study Comparing Humans and an Online Recommender System," *Proceedings of the Second ACM Conference on Recommender Systems—RecSys '08* (New York: ACM, 2008), 211–18.

27. See Ruth Towse, *A Textbook of Cultural Economics* (Cambridge: Cambridge University Press, 2010), 152.

28. Anindita Chakravarty, Yong Liu, and Tridib Mazumdar, "The Differential Effects of Online Word-of-Mouth and Critics' Reviews on Pre-release Movie Evaluation," *Journal of Interactive Marketing* 24, no. 3 (2010): 185–97; esp. 189.

29. David Wright, *Understanding Cultural Taste: Sensation, Skill and Sensibility* (Basingstoke, UK: Palgrave Macmillan, 2015), 153. The supposed loss of watercooler moments in the face of a more atomized media landscape has inspired much debate. See, for example, Elihu Katz, "And Deliver Us from Segmentation," *Annals of the American Academy of Political and Social Science* 546, no. 1 (1996): 22–33; Diana C. Mutz and Lori Young, "Communication and Public Opinion: Plus Ça Change?," *Public Opinions Quarterly* 75, no. 5 (2011): 1018–44; Amanda D. Lotz, *The Television Will Be Revolutionized*, 2nd rev. ed. (New York: New York University Press, 2014), esp. 32, 40–42, 263, 273; Chuck Tryon, *On-Demand Culture: Digital Delivery and the Future of Movies* (New Brunswick, NJ: Rutgers University Press, 2013), 177.

30. James G. Webster, *The Marketplace of Attention: How Audiences Take Shape in a Digital Age* (Cambridge, MA: MIT Press, 2014), 29–30, 137–38. Recently, Hanchard et al. have delivered an intriguing meta-analysis of two large, nationally representative data sets that supports the thesis that most film viewers (which they differentiate between "extensive omnivores," "basic omnivores," and "inactives") consume a wide variety of genres and that there are nine latent classes of film viewers who have a high probability of preferring and clustering around a subset of genres—not a single genre. See Matthew Hanchard et al., "Exploring Contemporary Patterns of Cultural Consumption: Offline and Online Film Watching in the UK," *Emerald Open Research* 1, no. 16 (2019).

31. Richard Maltby, "New Cinema Histories," in *Explorations in New Cinema History: Approaches and Case Studies*, ed. Richard Maltby, Daniel Biltereyst, and Philippe Meers (Chichester: Wiley-Blackwell, 2011), 9.

32. Annette Kuhn, "What to Do with Cinema Memory?," in *Explorations in New Cinema History: Approaches and Case Studies*, ed. Richard Maltby, Daniel Biltereyst, and Philippe Meers (Chichester: Wiley-Blackwell, 2011), 85.

33. Paul McEvoy, BFI Research and Statistics Unit, "Pick and Mix: Cultural Access, Screen Engagement and Film Appetites in the UK Regions," presentation at the online conference "Audiences beyond the Multiplex: Understanding the Value of a Diverse Film Culture," University of Glasgow, 2 March 2021. The hitherto two-wave (27 August—12 September 2019; 30 October—18 November 2020) nationally representative survey has an unweighted sample size of 12,029 UK adults.

34. Grant Blank, *Critics, Ratings, and Society: The Sociology of Reviews* (Lanham, MD: Rowman & Littlefield, 2007), 32, 158–59.

35. "Perceived similarity between the information source and the message recipient is an important determinant of source credibility, and hence persuasion." See Chakravarty, Liu, and Mazumdar, "The Differential Effects of Online Word-of-Mouth and Critics' Reviews on Pre-release Movie Evaluation," 189. Some research indicates that heavy users with narrow interests are more likely to be influenced by user forums, whereas users with wide interests (omnivores), tend to trust legacy institutions of criticism. See Marc Verboord, "The Legitimacy of Book Critics in the Age of the Internet and Omnivorousness: Expert Critics, Internet Critics and Peer Critics in Flanders and the Netherlands," *European Sociological Review* 26, no. 6 (2010): 623–37.

AFTERWORD: ROBOT CRITICS VS. HUMAN EXPERTS

1. See, for instance, Yochai Benkler, *The Wealth of Networks: How Social Production Transforms Markets and Freedom* (New Haven, CT: Yale University Press, 2006).

2. Neta Alexander, "Catered to Your Future Self: Netflix's 'Predictive Personalization' and the Mathematization of Taste," in *The Netflix Effect: Technology and Entertainment in the 21st Century*, ed. Kevin McDonald and Daniel Smith-Rowsey (New York: Bloomsbury, 2016), 93.

3. For a discussion of this phenomenon, see Amanda D. Lotz, *The Television Will Be Revolutionized*, 2nd rev. ed. (New York: New York University Press, 2014), 40–46. Concerns about an atomization of media audiences far precede the mass uptake of streaming (see Joseph Turow, *Breaking Up America: Advertisers and the New Media World* [Chicago: University of Chicago Press, 1997]). Likewise, they are not restricted to film, television, and media studies, as evidenced in work by Cass R. Sunstein and many others. Popular organs express these ideas in more pragmatic terms. See, for example, Kevin Roose, "The Messy, Confusing Future of TV? It's Here," *New York Times*, 13 August 2017, https://nyti.ms/2vvCsKe, which bemoans the broken promises of the "glorious, consumer-friendly future of TV" in which "incredible shows and movies would be a click away through low-cost, easy-to-use services," and the reality of a "hyper-fragmented mess, with a jumble of on-demand services."

4. European Commission, *A Profile of Current and Future Audiovisual Audience* (Luxembourg: Publications Office of the European Union, 2014), 45–58.

5. European Commission, *A Profile of Current and Future Audiovisual Audience*, 57–58.

6. European Audiovisual Observatory, *The Development of the European Market for On-Demand Audiovisual Services* (Strasbourg: European Audiovisual Observatory, 2015), 133, 148.

7. Emanuel Rosen, *The Anatomy of Buzz Revisited: Real-life Lessons in Word-of-Mouth Marketing* (New York: Doubleday, 2009), 245.

8. For the theoretical basis of this scenario, see Rosen, *The Anatomy of Buzz Revisited*, 139–41.

9. Vinod Krishnan et al., "Who Predicts Better?—Results from an Online Study Comparing Humans and an Online Recommender System," *Proceedings of the Second ACM Conference on Recommender Systems—RecSys '08* (New York: ACM, 2008), 211–18.

10. Kyung-Hyan Yoo, Ulrike Gretzel, and Markus Zander, "Source Factors in Recommender System Credibility Evaluation," in *Recommender Systems Handbook*, 2nd rev. ed., ed. Francesco Ricci, Lior Rokach, and Bracha Shapira (New York: Springer, 2015), esp. 691.

11. See Jonah Berger, *Contagious: How to Build Word of Mouth in the Digital Age* (London: Simon and Schuster, 2013), 7–9, 39.

12. Neil Hunt, quoted in Ken Auletta, "Outside the Box: Netflix and the Future of Television," *New Yorker*, 3 February 2014, https://www.newyorker.com/magazine /2014/02/03/outside-the-box-2.

13. In a June 2018 worldwide poll, only 23 percent of participants said they trusted Facebook. Matthew Moore, "The *Times* is Britain's Most Trusted Newspaper," *The Times* (London), 14 June 2018, 3.

14. On millennials' blasé attitudes toward Netflix's personalization, see Alison N. Novak, "Narrowcasting, Millennials and the Personalization of Genre in Digital Media," in *The Age of Netflix: Critical Essays on Streaming Media, Digital Delivery and Instant Access*, ed. Cory Barker and Myc Wiatrowski (Jefferson, NC: McFarland, 2017), 162–81.

15. Blake Hallinan and Ted Striphas, "Recommended for You: The Netflix Prize and the Production of Algorithmic Culture," *New Media and Society* 18, no. 1 (2014): 118. For another indicator of Netflix's naiveté in this respect, see Ted Sarandos's comments in "Ted Sarandos, Chief Content Officer, Netflix," in *Distribution Revolution: Conversations about the Digital Future of Film and Television*, ed. Michael Curtin, Jennifer Holt, and Kevin Sanson (Oakland: University of California Press, 2014), esp. 136.

16. Ramon Lobato, *Netflix Nations: The Geography of Digital Distribution* (New York: New York University Press, 2019), 183.

17. See Peter J. Rentfrow, Lewis R. Goldberg, and Ran Zilca, "Listening, Watching, and Reading: The Structure and Correlates of Entertainment Preferences," *Journal of Personality* 79, no. 2 (2011): 223–58; Marko Tkalcic and Li Chen, "Personality and Recommender Systems," in *Recommender Systems Handbook*, 2nd rev. ed., ed. Francesco Ricci, Lior Rokach, and Bracha Shapira (New York: Springer, 2015), 715–39.

18. See OfCom, *Communications Market Report 2017* (London: OfCom, 2017); European Commission, *A Profile of Current and Future Audiovisual Audience*; Elizabeth Evans and Paul McDonald, "Online Distribution of Film and Television in the UK: Behavior, Taste, and Value," in *Connected Viewing: Selling, Streaming, and Sharing Media in the Digital Era*, ed. Jennifer Holt and Kevin Sanson (London: Routledge, 2014).

19. See, for example, Raymond Williams, *Marxism and Literature* (Oxford: Oxford University Press, 1977), 121–27.

20. Douglas Gomery demonstrates how the emergence of television does not fully explain the decline of theatrical cinemagoing in the mid-twentieth century. See his *Shared Pleasures: A History of Movie Presentation in the United States* (Madison: University of Wisconsin Press, 1992), 83–85.

21. OfCom, *Media Nations: UK 2019* (London: OfCom, 2019), 60.

22. Webster explains the "law of double jeopardy" thus: "As a rule, it's heavy users who are in the audience for unpopular offerings. Popular offerings attract light users. That's what makes them popular." James G. Webster, *The Marketplace of Attention: How Audiences Take Shape in a Digital Age* (Cambridge, MA: MIT Press, 2014), 113. See also Motion Picture Association of America, *MPAA Theatrical Market Statistics 2016* (Los Angeles, CA: MPAA, 2016), 13.

23. See EY QUEST/National Association of Theatre Owners, *The Relationship between Movie Theater Attendance and Streaming Behavior* (Washington, DC: National Association of Theatre Owners, 2018); Huw D. Jones, "Watching Films On-Demand: A Report on Film Consumption on VOD Platforms in the UK," University of York, September 2017, https://www.academia.edu/34959355/Watching_films_on-demand_A_report_on_film_consumption_on_VOD_platforms_in_the_UK; OfCom, *Media Nations: UK 2019*.

24. James Cooray, "Predicted the Death of Broadcast TV? Then Have I Got News For You," *New Statesman*, 20 March 2017, https://www.newstatesman.com/culture/tv-radio/2017/03/predicted-death-broadcast-tv-then-have-i-got-news-you; OfCom, *Media Nations: UK 2019*.

25. For an example of early utopian predictions of "the entire surviving history of movies" being "open for browsing and sampling" via streaming, a utopian jukebox of the breadth of cultural production available at the click of a button and for a peppercorn fee, see A. O. Scott, "The Shape of Cinema, Transformed at the Click of a Mouse," *New York Times*, 18 March 2007, http://www.nytimes.com/2007/03/18/movies/18scot.html.

26. Karol Jakubowicz, "New Media Ecology: Reconceptualizing Media Pluralism," in *Media Pluralism and Diversity: Concepts, Risks and Global Trends*, ed. Peggy Valcke, Miklós Sükösd, and Robert G. Picard (Basingstoke, UK: Palgrave Macmillan, 2015), 42.

27. See, for example, Daniel G. Williams, *Ethnicity and Cultural Authority: From Arnold to Du Bois* (Edinburgh: Edinburgh University Press, 2006).

28. *Infomediaries* is a term from Jeremy Wade Morris, "Curation by Code: Infomediaries and the Data Mining of Taste," *European Journal of Cultural Studies* 18, nos. 4–5 (2015): 446–63.

29. See Dietmar Jannach et al., *Recommender Systems: An Introduction* (Cambridge: Cambridge University Press, 2011), 81–82, 291, 301.

30. Pierre Bourdieu, *Distinction: A Social Critique of the Judgment of Taste*, trans. Richard Nice (Cambridge, MA: Harvard University Press, 1984 [1979]), 234.

31. For an example of the long line of economics research demonstrating critics' insignificant effect on box-office performance, see David A. Reinstein and Christopher M. Snyder, "The Influence of Expert Reviews on Consumer Demand

for Experience Goods: A Case Study of Movie Critics," *Journal of Industrial Economics* 53, no. 1 (2005): 27–51. Numerous audience studies, conducted from the 1950s until as recently as 2014, confirm that whereas heavy users tend to engage with criticism, infrequent film consumers rely instead on word of mouth and advertisements. One recent example: European Commission, *A Profile of Current and Future Audiovisual Audience.*

32. See Mattias Frey, *The Permanent Crisis of Film Criticism: The Anxiety of Authority* (Amsterdam: Amsterdam University Press, 2015), e.g., 101–23.

33. See, for example, Rónan McDonald, *The Death of the Critic* (London: Continuum, 2007); cf. Frey, *The Permanent Crisis of Film Criticism.*

34. On reviewing's basic questions, see David Wright, *Understanding Cultural Taste: Sensation, Skill and Sensibility* (Basingstoke, UK: Palgrave Macmillan, 2015), 132.

35. Anthony Jameson et al., "Human Decision Making and Recommender Systems," in *Recommender Systems Handbook*, 2nd rev. ed., ed. Francesco Ricci, Lior Rokach, and Bracha Shapira (New York: Springer, 2015), 611.

36. See, for example, the account of newspapers' online ventures in Pablo J. Boczkowski, *Digitizing the News: Innovation in Online Newspapers* (Cambridge, MA: MIT Press, 2004), esp. 20, 42.

37. Jay David Bolter and Richard Grusin, *Remediation: Understanding New Media* (Cambridge, MA: MIT Press, 1999), 14–15.

38. For clearheaded assessments of this long-running saga, which was reawakened in the twentieth century by C. P. Snow's lecture on "The Two Cultures" and has continued, unabated and with increased intensity in the last twenty years, see, for example, Guy Ortalano, *The Two Cultures Controversy: Science, Literature and Cultural Politics in Postwar Britain* (Cambridge: Cambridge University Press, 2009); Murray Smith, *Film, Art, and the Third Culture: A Naturalized Aesthetics of Film* (Oxford: Oxford University Press, 2017). For more on the "paradigm wars," see Alan Bryman, *Social Research Methods*, 5th rev. ed. (Oxford: Oxford University Press, 2016), 657.

39. danah boyd and Kate Crawford, "Critical Questions for Big Data: Provocations for a Cultural, Technological, and Scholarly Phenomenon," *Information, Communication and Society* 15, no. 5 (2012): 667.

40. See Andrew McAfee, "Who are the Humanists and Why Do They Dislike Technology So Much?," *Financial Times*, 7 July 2015, https://www.ft.com/content/8fbd6859-def5-35ec-bbf5-0c262469e3e9; Chris Anderson, "The End of Theory: The Data Deluge Makes the Scientific Method Obsolete," *Wired*, 23 June 2008, https://www.wired.com/2008/06/pb-theory/.

41. Steven Pinker, "Science Is Not Your Enemy," *The New Republic*, 7 August 2013, https://newrepublic.com/article/114127/science-not-enemy-humanities.

42. Jack Nicas, "Apple's Radical Approach to News: Humans Over Machines," *New York Times*, 25 October 2018, https://www.nytimes.com/2018/10/25/technology/apple-news-humans-algorithms.html.

43. Shalini Ramachandran and Joe Flint, "At Netflix, Hollywood Battles the Algorithm," *Wall Street Journal*, 10 November 2018, B1.

44. Melanie Mitchell, "Artificial Intelligence Hits the Barrier of Meaning," *New York Times*, 5 November 2018, https://www.nytimes.com/2018/11/05/opinion/artificial-intelligence-machine-learning.html.

45. Helmut Martin-Jung, "Künstliche Intelligenz wird überschätzt," *Süddeutsche Zeitung*, 30 November 2018, https://www.sueddeutsche.de/digital/kuenstliche-intelligenz-digitalgipfel-regierung-algorithmen-1.4233675.

46. Quoted in Jonathan Shaw, "Artificial Intelligence and Ethics: Ethics and the Dawn of Decision-Making Machines," *Harvard Magazine*, January-February 2019, https://harvardmagazine.com/2019/01/artificial-intelligence-limitations.

47. See, for example, John Harris, "Meet Your New Cobot: Is a Machine Coming for Your Job?," *Guardian*, 25 November 2017, https://www.theguardian.com/money/2017/nov/25/cobot-machine-coming-job-robots-amazon-ocado; Peter S. Goodman, "The Robots Are Coming, and Sweden Is Fine," *New York Times*, 27 December 2017, https://www.nytimes.com/2017/12/27/business/the-robots-are-coming-and-sweden-is-fine.html. The term "useless class" derives from Yuval Noah Harari, *Homo Deus: A Brief History of Tomorrow* (London: Harvill Secker, 2016). See also Erik Brynjolfsson and Andrew McAfee, *Race against the Machine: How the Digital Revolution Is Accelerating Innovation, Driving Productivity, and Irreversibly Transforming Employment and the Economy* (Lexington, MA: Digital Frontier, 2011).

48. Richard Berriman and John Hawksworth, "Will Robots Steal Our Jobs? The Potential Impact of Automation on the United Kingdom and Other Economies," *PwC: UK Economic Outlook*, March 2017, https://www.pwc.co.uk/economic-services/ukeo/pwcukeo-section-4-automation-march-2017-v2.pdf.

49. Benedikt Frey and Michael A. Osborne, "The Future of Employment: How Susceptible Are Jobs to Computerisation?," *Working Paper of the Oxford Martin Programme on Technology and Employment*, 17 September 2013, https://www.oxfordmartin.ox.ac.uk/downloads/academic/The_Future_of_Employment.pdf.

50. Frey and Osborne, "The Future of Employment," 3.

51. Professor Ian Goldin, Oxford University, comments on *The Big Questions*, series 11, episode 17, "Could Robots and Artificial Intelligence Do More Harm Than Good?," aired 27 May 2018, on BBC1.

52. Michael Chui, James Manyika, and Mehdi Miremedi, "Where Machines Could Replace Humans—and Where They Can't (Yet)," *McKinsey Quarterly*, July 2016, https://www.mckinsey.com/business-functions/mckinsey-digital/our-insights/where-machines-could-replace-humans-and-where-they-cant-yet.

53. Frey and Osborne, "The Future of Employment," 18–19. For study on Israeli judges' human biases, see Shai Danziger, Jonathan Levav, and Liora Avnaim-Pesso, "Extraneous Factors in Judicial Decisions," *Proceedings of the National Academy of Sciences* 108, no. 17 (2001): 6889–92. For a techno-optimistic account of using algorithms to detect children in danger and reduce unconscious race bias in the

Pittsburgh child protection agency, see Dan Hurley, "Can an Algorithm Tell When Kids Are in Danger?," *New York Times*, 2 January 2018, https://nyti.ms/2D002Xc.

54. Christopher Steiner, *Automate This: How Algorithms Took Over Our Markets, Our Jobs, and the World* (New York: Portfolio/Penguin, 2012); Alex Rosenblat, *Uberland: How Algorithms Are Rewriting the Rules of Work* (Oakland: University of California Press, 2018).

55. Steiner, *Automate This*, 5.

56. Aaron Smith and Monica Anderson, *Automation in Everyday Life* (Washington, DC: Pew Research Center, 2017).

57. Simon Romero, "Wielding Rocks and Knives, Arizonans Attack Self-Driving Cars," *New York Times*, 31 December 2018, https://www.nytimes.com/2018/12/31/us/waymo-self-driving-cars-arizona-attacks.html.

58. Michael Savage, "Rise of Robots 'Could See Workers Enjoy Four-Day Weeks,'" *Guardian*, 13 October 2018, https://www.theguardian.com/technology/2018/oct/13/rise-robots-four-day-working-week.

59. Jaclyn Peiser, "The Rise of the Robot Reporter," *New York Times*, 5 February 2019, https://www.nytimes.com/2019/02/05/business/media/artificial-intelligence-journalism-robots.html; Goodman, "The Robots Are Coming."

60. Lucy Bannerman, "Family Portraits Put the Art into Artificial Intelligence," *The Times* (London), 23 August 2018, 19.

61. Louis Anslow, "Robots Have Been About to Take All the Jobs for More than 200 Years," *Timeline*, 16 May 2016, https://timeline.com/robots-have-been-about-to-take-all-the-jobs-for-more-than-200-years-5c9c08a2f41d.

62. Paul Krugman, "Democrats, Avoid the Robot Rabbit Hole," *New York Times*, 17 October 2019, https://www.nytimes.com/2019/10/17/opinion/democrats-automation.html.

63. William Uricchio, "Data, Culture and the Ambivalence of Algorithms," in *The Datafied Society: Studying Culture through Data*, ed. Mirko Tobias Schäfer and Karin van Es (Amsterdam: Amsterdam University Press, 2017), 128.

64. Tarleton Gillespie, "The Relevance of Algorithms," in *Media Technologies: Essays on Communication, Materiality, and Society*, ed. Tarleton Gillespie, Pablo J. Boczkowski, and Kirsten A. Foot (Cambridge, MA: MIT Press, 2014), 169.

65. For a discussion of newspapers' flawed response to the internet, see Boczkowski, *Digitizing the News*, 20.

66. See, for example, Walter Benjamin, "The Work of Art in the Age of Mechanical Reproduction," trans. Harry Zohn, in *Illuminations*, ed. Hannah Arendt (New York: Schocken, 1969 [1936]), 217–51; Henry Jenkins, *Convergence Culture: Where Old and New Media Collide*, 2nd rev. ed. (New York: New York University Press, 2006), 13–16.

67. On distinctions between reviewing and criticism, see Wesley Monroe Shrum Jr., *Fringe and Fortune: The Role of Critics in High and Popular Art* (Princeton, NJ: Princeton University Press, 1996), 44. For a deliberation on the "like" economy, see Carolin Gerlitz and Anne Helmon, "The Like Economy: Social Buttons and the Data-Intensive Web," *New Media and Society* 15, no. 8 (2013): 1348–65.

68. Lisa Gitelman, *Always Already New: Media, History, and the Data of Culture* (Cambridge, MA: MIT Press, 2006), 6.

APPENDIX: DESIGNING THE EMPIRICAL
AUDIENCE STUDY

1. European Commission, *A Profile of Current and Future Audiovisual Audience* (Luxembourg: Publications Office of the European Union, 2014).

2. Hennink, Kaiser, and Marconi usefully articulate the difference between the two: data saturation or "code saturation may indicate when researchers have 'heard it all,' but meaning [or theoretical saturation] is needed to 'understand it all.'" Data saturation "refers to the point in data collection when no additional issues are identified, data begin to repeat, and further data collection becomes redundant," whereas theoretical saturation "refers to the point in data collection when no additional issues or insights emerge from data and all relevant conceptual categories have been identified, explored, and exhausted." See Monique M. Hennink, Bonnie N. Kaiser, and Vincent C. Marconi, "Code Saturation Versus Meaning Saturation: How Many Interviews Are Enough?," *Qualitative Health Research* 27, no. 4 (2017): 591–92. On this question see also: Ashley K. Hagaman and Amber Wutich, "How Many Interviews Are Enough to Identify Metathemes in Multisited and Cross-Cultural Research? Another Perspective on Guest, Bunce, and Johnson's (2006) Landmark Study," *Field Methods* 29, no. 1 (2017): 23–41; Alan Bryman, *Social Research Methods*, 5th rev. ed. (Oxford: Oxford University Press, 2016), 410–18.

SELECTED BIBLIOGRAPHY

This bibliography contains the most important monographs and academic journal articles. For reasons of space and because they are included in the notes, statistical reports, conference proceedings, business-school case studies, working papers, interviews, and trade and newspaper articles are omitted here.

Alexander, Neta. "Catered to Your Future Self: Netflix's 'Predictive Personalization' and the Mathematization of Taste." In *The Netflix Effect: Technology and Entertainment in the 21st Century*, edited by Kevin McDonald and Daniel Smith-Rowsey, 81–97. New York: Bloomsbury, 2016.

Anderson, Chris. *The Longer Long Tail: How Endless Choice Is Creating Unlimited Demand*. 2nd rev. ed. London: Random House, 2009.

Anderson, Eugene W. "Consumer Satisfaction and Word of Mouth." *Journal of Services Research* 1, no. 1 (1998): 5–17.

Arnold, Matthew. *Culture and Anarchy and Other Writings*. Edited by Stefan Collini. Cambridge: Cambridge University Press, 1993 [1869].

Arnold, Sarah. "Netflix and the Myth of Choice/Participation/Autonomy." In *The Netflix Effect: Technology and Entertainment in the 21st Century*, edited by Kevin McDonald and Daniel Smith-Rowsey, 49–62. New York: Bloomsbury, 2016.

Atkinson, Sarah. *Beyond the Screen: Emerging Cinema and Engaging Audiences*. New York: Bloomsbury, 2014.

Austin, Bruce A. "Film Attendance: Why College Students Chose to See Their Most Recent Film." *Journal of Popular Film and Television* 9, no. 1 (1982): 43–49.

———. "Rating the Movies." *Journal of Popular Film and Television* 7, no. 4 (1980): 384–99.

Balanzategui, Jessica, Liam Burke, and Dan Golding. "Recommending a New System: An Audience-Based Approach to Film Categorisation in the Digital Age." *Participations* 15, no. 2 (2018): 297–328.

Basuroy, Suman, Subimal Chatterjee, and S. Abraham Ravid. "How Critical Are Critical Reviews? The Box Office Effects of Film Critics, Star Power, and Budgets." *Journal of Marketing* 67, no. 4 (2003): 103–17.

Baumann, Shyon. *Hollywood Highbrow: From Entertainment to Art*. Princeton, NJ: Princeton University Press, 2007.

Beer, David. *Popular Culture and New Media: The Politics of Circulation*. Basingstoke, UK: Palgrave Macmillan, 2013.

Benjamin, Walter. "The Work of Art in the Age of Mechanical Reproduction." Translated by Harry Zohn. In *Illuminations*, edited by Hannah Arendt, 217–51. New York: Schocken, 1969 [1936].

Benkler, Yochai. *The Wealth of Networks: How Social Production Transforms Markets and Freedom*. New Haven, CT: Yale University Press, 2006.

Berger, Jonah. *Contagious: How to Build Word of Mouth in the Digital Age*. London: Simon and Schuster, 2013.

Bettman, James R., Mary Francis Luce, and John W. Payne. "Constructive Consumer Choice Processes." *Journal of Consumer Research* 25, no. 3 (1998): 187–217.

Blank, Grant. *Critics, Ratings, and Society: The Sociology of Reviews*. Lanham, MD: Rowman & Littlefield, 2007.

Boatwright, Peter, Suman Basuroy, and Wagner A. Kamakura. "Reviewing the Reviewers: The Impact of Individual Film Critics on Box Office Performance." *Quantitative Marketing and Economics* 5, no. 4 (2007): 401–25.

Boczkowski, Pablo J. *Digitizing the News: Innovation in Online Newspapers*. Cambridge, MA: MIT Press, 2004.

Bolter, Jay David, and Richard Grusin. *Remediation: Understanding New Media*. Cambridge, MA: MIT Press, 1999.

Bostrom, Nick. *Superintelligence: Paths, Dangers, Strategies*. Oxford: Oxford University Press, 2014.

Bourdieu, Pierre. *Distinction: A Social Critique of the Judgment of Taste*. Translated by Richard Nice. London: Routledge, 1984.

boyd, danah, and Kate Crawford. "Critical Questions for Big Data: Provocations for a Cultural, Technological, and Scholarly Phenomenon." *Information, Communication and Society* 15, no. 5 (2012): 662–79.

Brown, Jacqueline Johnson, and Peter H. Reingen. "Social Ties and Word-of-Mouth Referral Behavior." *Journal of Consumer Research* 14, no. 3 (1987): 350–62.

Brown, Jo, Amanda J. Broderick, and Nick Lee. "Word of Mouth Communication within Online Communities: Conceptualizing the Online Social Network." *Journal of Interactive Marketing* 21, no. 3 (2007): 2–20.

Brynjolfsson, Erik, and Andrew McAfee. *Race against the Machine: How the Digital Revolution Is Accelerating Innovation, Driving Productivity, and Irreversibly Transforming Employment and the Economy*. Lexington, MA: Digital Frontier, 2011.

Carroll, Noël. *On Criticism*. New York: Routledge, 2009.

Castells, Pablo, Neil J. Hurley, and Saul Vargas. "Novelty and Diversity in Recommender Systems." In *Recommender Systems Handbook*, 2nd rev. ed., edited by Francesco Ricci, Lior Rokach, and Bracha Shapira, 881–918. New York: Springer, 2015.

Chabert, Jean-Luc, ed. *A History of Algorithms: From the Pebble to the Microchip*. Translated by Chris Weeks. Berlin: Springer, 1999.

Chakravarty, Anindita, Yong Liu, and Tridib Mazumdar. "The Differential Effects of Online Word-of-Mouth and Critics' Reviews on Pre-release Movie Evaluation." *Journal of Interactive Marketing* 24, no. 3 (2010): 185–97.

Chen, Meng-Hui, Chin-Hung Teng, and Pei-Chann Chang. "Applying Artificial Immune Systems to Collaborative Filtering for Movie Recommendation." *Advanced Engineering Informatics* 29, no. 4 (2015): 830–39.

Cheney-Lippold, John. "A New Algorithmic Identity: Soft Biopolitics and the Modulation of Control." *Theory, Culture and Society* 28, no. 6 (2011): 164–81.

———. *We Are Data: Algorithms and the Making of Our Digital Selves.* New York: New York University Press, 2017.

Cohn, Jonathan. *The Burden of Choice: Recommendations, Subversion, and Algorithmic Culture.* New Brunswick, NJ: Rutgers University Press, 2019.

Corliss, Richard. "All Thumbs or, Is There a Future for Film Criticism?" *Film Comment* 26, no. 2 (1990): 14–18.

Cunningham, Stuart, and Jon Silver. *Screen Distribution and the New King Kongs of the Online World.* Basingstoke, UK: Palgrave Macmillan, 2013.

Curtin, Michael, Jennifer Holt, and Kevin Sanson, eds. *Distribution Revolution: Conversations about the Digital Future of Film and Television.* Oakland: University of California Press, 2014.

Deuchert, Eva, Kossi Adjamah, and Florian Pauly. "For Oscar Glory or Oscar Money? Academy Awards and Movie Success." *Journal of Cultural Economics* 29, no. 3 (2005): 159–76.

De Valck, Marijke. "Convergence, Digitisation and the Future of Film Festivals." In *Digital Disruption: Cinema Moves On-line*, edited by Dina Iordanova and Stuart Cunningham, 117–29. St. Andrews, UK: St. Andrews Film Studies, 2012.

De Vany, Arthur S., and W. David Walls. "Bose-Einstein Dynamics and Adaptive Contracting in the Motion Picture Industry." *Economic Journal* 106, no. 439 (1996): 1493–514.

Dodds, John C., and Morris B. Holbrook. "What's an Oscar Worth? An Empirical Estimation of the Effects of Nominations and Awards on Movie Distribution and Revenues." In *Current Research in Film: Audiences, Economics, and Law.* Vol. 4, edited by Bruce A. Austin, 72–88. Norwood, NJ: Abex, 1988.

Domingos, Pedro. *The Master Algorithm: How the Quest for the Ultimate Learning Machine Will Remake Our World.* New York: Basic Books, 2015.

Dourish, Paul. "Algorithms and Their Others: Algorithmic Culture in Context." *Big Data and Society* 3, no. 2 (2016): 1–11.

Duan, Wenjing, Bin Gu, and Andrew Whinston. "Do Online Reviews Matter?—An Empirical Investigation of Panel Data." *Decision Support Systems* 45, no. 4 (2008): 1007–16.

East, Robert, Kathy Hammond, and Malcolm Wright. "The Relative Incidence of Positive and Negative Word of Mouth: A Multi-Category Study." *International Journal of Research and Marketing* 24, no. 2 (2007): 175–84.

Eliashberg, Jehoshua, and Steven M. Shugan. "Film Critics: Influencers or Predictors?" *Journal of Marketing* 61, no. 2 (1997): 68–78.

Elmer, Greg. *Profiling Machines: Mapping the Personal Information Economy.* Cambridge, MA: MIT Press, 2004.

English, James F. "The Economics of Cultural Awards." In *Handbook of the Economics of Art and Culture.* Vol. 2, edited by Victoria A. Ginsburgh and David Throsby, 119–43. Oxford: North-Holland, 2014.

Ensmenger, Nathan. *The Computer Boys Take Over: Computers, Programmers, and the Politics of Technical Expertise.* Cambridge, MA: MIT Press, 2010.

Evans, Elizabeth, and Paul McDonald. "Online Distribution of Film and Television in the UK: Behavior, Taste, and Value." In *Connected Viewing: Selling, Streaming, and Sharing Media in the Digital Era*, edited by Jennifer Holt and Kevin Sanson, 158–79. New York: Routledge, 2014.

Ewen, Stuart. *Captains of Consciousness: Advertising and the Social Roots of the Consumer Culture.* 2nd rev. ed. New York: Basic Books, 2001.

Faber, Ronald J., and Thomas C. O'Guinn. "Effect of Media Advertising and Other Sources on Movie Selection." *Journalism Quarterly* 61, no. 2 (1984): 371–77.

Faber, Ronald J., Thomas C. O'Guinn, and Andrew P. Hardy. "Art Films in the Suburbs: A Comparison of Popular and Art Film Audiences." In *Current Research in Film: Audiences, Economics, and Law.* Vol. 4, edited by Bruce A. Austin, 45–53. Norwood, NJ: Ablex, 1988.

Farchy, Joëlle. "Die Bedeutung von Information für die Nachfrage nach kulturellen Gütern." Translated by Vinzenz Hediger. In *Demnächst in Ihrem Kino: Grundlagen der Filmwerbung und Filmvermarktung*, edited by Vinzenz Hediger and Patrick Vonderau, 193-211. Marburg: Schüren, 2005.

Franck, Georg. Ökonomie der Aufmerksamkeit: Ein Entwurf. Munich: Carl Hanser, 1998.

Frank, Robert H., and Philip J. Cook. *The Winner-Take-All Society: Why the Few at the Top Get So Much More Than the Rest of Us.* 2nd rev. ed. London: Virgin, 2010.

Freeman, Matthew. *Industrial Approaches to Media: A Methodological Gateway to Industry Studies.* London: Palgrave Macmillan, 2016.

Frey, Mattias. "Critical Questions." In *Film Criticism in the Digital Age*, edited by Mattias Frey and Cecilia Sayad, 1-20. New Brunswick, NJ: Rutgers University Press, 2015.

———. "The Ends of (German) Film Criticism: On Recurring Doomsday Scenarios and the New Algorithmic Culture." *New German Critique* 47, no. 3 (2020): 45–57.

———. *Extreme Cinema: The Transgressive Rhetoric of Today's Art Cinema Culture.* New Brunswick, NJ: Rutgers University Press, 2016.

———. "The Internet Suggests: Film, Recommender Systems, and Cultural Mediation." *Journal of Cinema and Media Studies* 59, no. 1 (2019): 163–69.

———. *MUBI and the Curation Model of Video on Demand.* London: Palgrave Macmillan, forthcoming.

———. *The Permanent Crisis of Film Criticism: The Anxiety of Authority.* Amsterdam: Amsterdam University Press, 2015.

Geiger, Theodor. "A Radio Taste of Musical Taste." *Public Opinion Quarterly* 14, no. 3 (1950): 453–60.

Gerlitz, Carolin, and Anne Helmon. "The Like Economy: Social Buttons and the Data-Intensive Web." *New Media and Society* 15, no. 8 (2013): 1348–65.

Gigerenzer, Gerd, and Reinhard Selten, eds. *Bounded Rationality: The Adaptive Toolbox.* Cambridge, MA: MIT Press, 2001.

Gillespie, Tarleton. "The Relevance of Algorithms." In *Media Technologies: Essays on Communication, Materiality, and Society*, edited by Tarleton Gillespie, Pablo J. Boczkowski, and Kirsten A. Foot, 167–93. Cambridge, MA: MIT Press, 2014.

Gitelman, Lisa. *Always Already New: Media, History, and the Data of Culture.* Cambridge, MA: MIT Press, 2006.

Gitlin, Todd. "Public Sphere or Public Sphericles?" In *Media, Ritual, Identity*, edited by Tamar Liebes and James Curran, 168–75. London: Routledge, 1998.

Gomery, Douglas. *Shared Pleasures: A History of Movie Presentation in the United States.* Madison: University of Wisconsin Press, 1992.

Gomez-Uribe, Carlos A., and Neil Hunt. "The Netflix Recommender System: Algorithms, Business Value, and Innovation." *ACM Transactions on Management Information Systems* 6, no. 4, article 13 (2015): 1–19.

Gubbins, Michael. "Digital Revolution: Active Audiences and Fragmented Consumption." In *Digital Disruption: Cinema Moves On-Line*, edited by Dina Iordanova and Stuart Cunningham, 67–100. St. Andrews, UK: St. Andrews Film Studies, 2012.

Gunter, Barrie. *Predicting Success at the Box Office.* London: Palgrave Macmillan, 2018.

Hallinan, Blake, and Ted Striphas. "Recommended for You: The Netflix Prize and the Production of Algorithmic Culture." *New Media and Society* 18, no. 1 (2014): 117–37.

Hanchard, Matthew, Peter Merrington, Bridgette Wessels, and Simeon Yates. "Exploring Contemporary Patterns of Cultural Consumption: Offline and Online Film Watching in the UK." *Emerald Open Research* 1, no. 16 (2019): 1–24.

Harari, Yuval Noah. *Homo Deus: A Brief History of Tomorrow.* London: Harvill Secker, 2016.

Hasebrink, Uwe, and Hanna Domeyer. "Media Repertoires as Patterns of Behaviour and as Meaningful Practices: A Multimethod Approach to Media Use in Converging Media Environments." *Participations* 9, no. 2 (2012): 757–79.

Havens, Timothy, Amanda D. Lotz, and Serra Tinic. "Critical Media Industry Studies: A Research Approach." *Communication, Culture and Critique* 2, no. 2 (2009): 234–53.

Hediger, Vinzenz. "Die Werbung hat das erste Wort, der Zuschauer das letzte: Filmwerbung und das Problem der symmetrischen Ignoranz." In *Strategisches Management für Film- und Fernsehproduktion: Herausforderungen, Optionen, Kompetenzen*, edited by Michael Hülsmann and Jörn Grapp, 535–52. Munich: Oldenbourg Verlag, 2009.

Hennig-Thurau, Thorsten, Kevin P. Gwinner, Gianfranco Walsh, and Dwayne D. Gremler. "Electronic Word-of-Mouth via Consumer-Opinion Platforms: What Motivates Consumers to Articulate Themselves on the Internet." *Journal of Interactive Marketing* 18, no. 1 (2004): 38–52.

Hennig-Thurau, Thorsten, Caroline Wiertz, and Fabian Feldhaus. "Does Twitter Matter? The Impact of Microblogging Word of Mouth on Consumers' Adoption of New Movies." *Journal of the Academy of Marketing Science* 43, no. 3 (2015): 375–94.

Herbert, Daniel. *Videoland: Movie Culture at the American Video Store*. Berkeley: University of California Press, 2014.

Herder, Johann Gottfried. *Selected Writings on Aesthetics*. Edited and translated by Gregory Moore. Princeton, NJ: Princeton University Press, 2006.

Hilgartner, Stephen. *Science on Stage: Expert Advice as Public Drama*. Stanford, CA: Stanford University Press, 2000.

Hirsch, Paul M. "Processing Fads and Fashions: An Organization-Set Analysis of Cultural Industry Systems." *American Journal of Sociology* 77, no. 4 (1972): 639–59.

Hirst, Martin, and John Harrison. *Communication and New Media: From Broadcast to Narrowcast*. Oxford: Oxford University Press, 2007.

Holbrook, Morris B. "Popular Appeal versus Expert Judgments of Motion Pictures." *Journal of Consumer Research* 26, no. 2 (1999): 144–55.

Holt, Jennifer, and Alisa Perren, eds. *Media Industries: History, Theory, and Method*. Chichester, UK: Wiley-Blackwell, 2009.

Holt, Jennifer, and Kevin Sanson, eds. *Connected Viewing: Selling, Streaming and Sharing Media in the Digital Era*. New York: Routledge, 2014.

Hsu, Greta, and Joel M. Podolny. "Critiquing the Critics: An Approach for the Comparative Evaluation of Critical Schemas." *Social Science Research* 34, no. 1 (2005): 189–214.

Huhtamo, Erkki, and Jussi Parikka, eds. *Media Archaeology: Approaches, Applications, and Implications*. Berkeley: University of California Press, 2011.

Hume, David. *Four Dissertations*. London: A. Millar, 1757.

Jacobs, Ruud S., Ard Heuvelman, Somaya Ben Allouch, and Oscar Peters. "Everyone's a Critic: The Power of Expert and Consumer Reviews to Shape Readers' Postviewing Motion Picture Evaluations." *Poetics*, no. 52 (2015): 91–103.

Jakubowicz, Karol. "New Media Ecology: Reconceptualizing Media Pluralism." In *Media Pluralism and Diversity: Concepts, Risks and Global Trends*, edited by Peggy Valcke, Miklós Sükösd, and Robert G. Picard, 23–53. Basingstoke, UK: Palgrave Macmillan, 2015.

Jameson, Anthony, Martijn C. Willemsen, Alexander Felfernig, Marco de Gemmis, Pasquale Lops, Giovanni Semeraro, and Li Chen. "Human Decision Making and Recommender Systems." In *Recommender Systems Handbook*, 2nd rev. ed., edited by Francesco Ricci, Lior Rokach, and Bracha Shapira, 611–48. New York: Springer, 2015.

Jannach, Dietmar, Markus Zanker, Alexander Felfernig, and Gerhard Friedrich. *Recommender Systems: An Introduction*. Cambridge: Cambridge University Press, 2011.

Jenkins, Henry. *Convergence Culture: Where Old and New Media Collide*. 2nd rev. ed. New York: New York University Press, 2006.

Jenner, Mareike. "Binge-Watching: Video-on-Demand, Quality TV and Mainstreaming Fandom." *International Journal of Cultural Studies* 20, no. 3 (2017): 304–20.

———. *Netflix and the Re-invention of Television*. Basingstoke, UK: Palgrave, 2018.

Johnson, Catherine. *Online TV*. London: Routledge, 2019.

Kant, Immanuel. *Critique of Judgment*. Edited and translated by J. H. Bernard. London: Macmillan, 1914.

Katz, Elihu. "And Deliver Us from Segmentation." *Annals of the American Academy of Political and Social Science* 546, no. 1 (1996): 22–33.

Keating, Gina. *Netflixed: The Epic Battle for America's Eyeballs*. 2nd rev. ed. New York: Portfolio/Penguin, 2013.

Kerrigan, Finola. *Film Marketing*. 2nd rev. ed. Abingdon, UK: Routledge, 2017.

Koren, Yehuda, and Robert Bell. "Advances in Collaborative Filtering." In *Recommender Systems Handbook*, 2nd rev. ed., edited by Francesco Ricci, Lior Rokach, and Bracha Shapira, 77–93. New York: Springer, 2015.

Koren, Yehuda, Robert Bell, and Chris Volinsky. "Matrix Factorization for Recommender Systems." *Computer*, no. 8 (2009): 30–37.

Kuhn, Annette. "What to Do with Cinema Memory?" In *Explorations in New Cinema History: Approaches and Case Studies*, edited by Richard Maltby, Daniel Biltereyst, and Philippe Meers, 85–97. Chichester: Wiley-Blackwell, 2011.

Lanham, Richard A. *The Economics of Attention: Style and Substance in the Age of Information*. Chicago: University of Chicago Press, 2006.

Lévy, Pierre. *Collective Intelligence: Mankind's Emerging World in Cyberspace*. Cambridge, MA: Perseus Books, 1997.

Lippmann, Walter. *Public Opinion*. New York: Macmillan, 1922.

Lobato, Ramon. *Netflix Nations: The Geography of Digital Distribution*. New York: New York University Press, 2019.

Lotz, Amanda D. *Portals: A Treatise on Internet-Distributed Television*. Ann Arbor, MI: Maize Books, 2017.

———. *The Television Will Be Revolutionized*. 2nd rev. ed. New York: New York University Press, 2014.

Lovink, Geert. *My First Recession: Critical Internet Culture in Transition*. Rotterdam: V2_/NAi, 2003.

Maier, Andrea, Zata Vickers, and J. Jeffrey Inman. "Sensory-Specific Satiety, Its Crossovers, and Subsequent Choice of Potato Chip Flavors." *Appetite* 49, no. 2 (2007): 419–28.

Maltby, Richard. "New Cinema Histories." In *Explorations in New Cinema History: Approaches and Case Studies*, edited by Richard Maltby, Daniel Biltereyst, and Philippe Meers, 3–40. Chichester: Wiley-Blackwell, 2011.

Manovich, Lev. "Can We Think without Categories?" *Digital Culture and Society* 4, no. 1 (2018): 17–28.

Markham, Annette, Simona Stavrova, and Max Schlüter. "Netflix, Imagined Affordances, and the Illusion of Control." *Netflix at the Nexus: Content, Practice, and Production in the Age of Streaming Television*, edited by Amber M. Buck and Theo Plothe, 29–46. Bern: Peter Lang, 2019.

McDonald, Rónan. *The Death of the Critic.* London: Continuum, 2007.

McMurria, John. *Republic on the Wire: Cable Television, Pluralism, and the Politics of New Technologies, 1948–1984.* New Brunswick, NJ: Rutgers University Press, 2017.

Metzger, Miriam J., Andrew J. Flanagin, and Ryan B. Medders. "Social and Heuristic Approaches to Credibility Evaluation Online." *Journal of Communication* 60, no. 3 (2010): 413–39.

Moeran, Brian, and Jesper Strandgaard Pedersen, eds. *Negotiating Values in the Creative Industries: Fairs, Festivals and Competitive Events.* Cambridge: Cambridge University Press, 2011.

Morozov, Evgeny. "Opposing the Exceptionalism of the Algorithm." In *The Datafied Society: Studying Culture through Data*, edited by Mirko Tobias Schäfer and Karin van Es, 245–48. Amsterdam: Amsterdam University Press, 2017.

Morris, Jeremy Wade. "Curation by Code: Infomediaries and the Data Mining of Taste." *European Journal of Cultural Studies* 18, nos. 4–5 (2015): 446–63.

Mosco, Vincent. *The Digital Sublime: Myth, Power, and Cyberspace.* Cambridge, MA: MIT Press, 2004.

Mudambi, Susan M., and David Schuff. "What Makes a Helpful Online Review? A Study of Customer Reviews on Amazon.com." *MIS Quarterly* 34, no. 1 (2010): 185–200.

Mutz, Diana C., and Lori Young. "Communication and Public Opinion: Plus Ça Change?" *Public Opinions Quarterly* 75, no. 5 (2011): 1018–44.

Negroponte, Nicholas. *Being Digital.* 2nd rev. ed. New York: Vintage, 1996.

Nelson, Elissa. "Windows into the Digital World: Distributor Strategies and Consumer Choice in an Era of Connected Viewing." In *Connected Viewing: Selling, Streaming, and Sharing Media in the Digital Era*, edited by Jennifer Holt and Kevin Sanson, 62–78. New York: Routledge, 2014.

Nelson, Phillip. "Information and Consumer Behavior." *Journal of Political Economy* 78, no. 2 (1970): 311–29.

Ning, Xia, Christian Desrosiers, and George Karypis. "A Comprehensive Survey of Neighborhood-Based Recommendation Methods." In *Recommender Systems Handbook*, 2nd rev. ed., edited by Francesco Ricci, Lior Rokach, and Bracha Shapira, 37–76. New York: Springer, 2015.

Noble, Safiya Umoja. *Algorithms of Oppression: How Search Engines Reinforce Racism.* New York: New York University Press, 2018.

Novak, Alison N. "Narrowcasting, Millennials and the Personalization of Genre in Digital Media." In *The Age of Netflix: Critical Essays on Streaming Media, Digital Delivery and Instant Access*, edited by Cory Barker and Myc Wiatrowski, 162–81. Jefferson, NC: McFarland, 2017.

O'Keefe, Daniel J. *Persuasion: Theory and Research*. 3rd rev. ed. Los Angeles: Sage, 2016.

O'Neil, Cathy. *Weapons of Math Destruction: How Big Data Increases Inequality and Threatens Democracy*. New York: Penguin Random House, 2016.

Ortalano, Guy. *The Two Cultures Controversy: Science, Literature and Cultural Politics in Postwar Britain*. Cambridge: Cambridge University Press, 2009.

Pariser, Eli. *The Filter Bubble: What the Internet Is Hiding from You*. London: Penguin, 2011.

Pasquale, Frank. *The Black Box Society: The Secret Algorithms That Control Money and Information*. Cambridge, MA: Harvard University Press, 2015.

Paßmann, Johannes, and Asher Boersma. "Unknowing Algorithms: On Transparency of Unopenable Black Boxes." In *The Datafied Society: Studying Culture through Data*, edited by Mirko Tobias Schäfer and Karin van Es, 139–46. Amsterdam: Amsterdam University Press, 2017.

Peterson, Richard, and Roger Kern. "Changing Highbrow Taste: From Snob to Omnivore." *American Sociological Review* 61, no. 5 (1996): 900–907.

Pilipets, Elena. "From Netflix Streaming to Netflix and Chill: The (Dis)connected Body of Serial Binge-Viewer." *Social Media and Society* 5, no. 4 (2019): 1–19.

Randolph, Marc. *That Will Never Work: The Birth of Netflix and the Amazing Life of an Idea*. London: Endeavour, 2019.

Reinstein, David A., and Christopher M. Snyder. "The Influence of Expert Reviews on Consumer Demand for Experience Goods: A Case Study of Movie Critics." *Journal of Industrial Economics* 53, no. 1 (2005): 27–51.

Rentfrow, Peter J., Lewis R. Goldberg, and Ran Zilca. "Listening, Watching, and Reading: The Structure and Correlates of Entertainment Preferences." *Journal of Personality* 79, no. 2 (2011): 223–58.

Ricci, Francesco, Lior Rokach, and Bracha Shapira, eds. *Recommender Systems Handbook*. 2nd rev. ed. New York: Springer, 2015.

Rieder, Bernhard. *Engines of Order: A Mechanology of Algorithmic Techniques*. Amsterdam: Amsterdam University Press, 2020.

Rosen, Emanuel. *The Anatomy of Buzz Revisited: Real-life Lessons in Word-of-Mouth Marketing*. New York: Doubleday, 2009.

Rosenblat, Alex. *Uberland: How Algorithms Are Rewriting the Rules of Work*. Oakland: University of California Press, 2018.

Rubenking, Bridget, and Cheryl Campanella Bracken. *Binge Watching: Motivations and Implications of Our Changing Viewing Behaviors*. Bern: Peter Lang, 2020.

Schwartz, Barry. *The Paradox of Choice: Why More Is Less*. New York: Ecco, 2004.

Schwartz, Barry, Andrew Ward, John Monterosso, Sonja Lyubomirsky, Katherine White, and Darrin R. Lehman. "Maximizing versus Satisficing: Happiness Is a Matter of Choice." *Journal of Personality and Social Psychology* 83, no. 5 (2002): 1178–97.

Segijn, Claire M., and Anastasia Kononova. "Audience, Media and Cultural Factors as Predictors of Multiscreen Use: A Comparative Study of the Netherlands and the United States." *International Journal of Communication*, no. 12 (2018): 4708–30.

Shrum, Wesley Monroe Jr. *Fringe and Fortune: The Role of Critics in High and Popular Art*. Princeton, NJ: Princeton University Press, 1996.

Simon, Herbert A. "A Behavioral Model of Rational Choice." *Quarterly Journal of Economics* 69, no. 1 (1955): 99–118.

Simonton, Dean Keith. "Cinematic Success Criteria and Their Predictors: The Art and Business of the Film Industry." *Psychology and Marketing* 26, no. 5 (2009): 400–420.

Smith, Michael D., and Rahul Telang. *Streaming, Sharing, Stealing: Big Data and the Future of Entertainment*. Cambridge, MA: MIT Press, 2016.

Smith-Rowsey, Daniel. "Imaginative Indices and Deceptive Domains: How Netflix's Categories and Genres Redefine the Long Tail." In *The Netflix Effect: Technology and Entertainment in the 21st Century*, edited by Kevin McDonald and Daniel Smith-Rowsey, 63–79. New York: Bloomsbury, 2016.

Sparviero, Sergio. "Hollywood Creative Accounting: The Success Rate of Major Motion Pictures." *Media Industries* 2, no. 1 (2015): 19–36.

Spector, Robert. *Amazon.com: Get Big Fast*. New York: HarperBusiness, 2000.

Steiner, Christopher. *Automate This: How Algorithms Took Over Our Markets, Our Jobs, and the World*. New York: Portfolio/Penguin, 2012.

Steiner, Emil. "Binge-Watching in Practice: The Rituals, Motives and Feelings of Streaming Video Viewers." In *The Age of Netflix: Critical Essays on Streaming Media, Digital Delivery and Instant Access*, edited by Cory Barker and Myc Wiatrowski, 141–61. Jefferson, NC: McFarland, 2017.

Steiner, Emil, and Kun Xu. "Binge-Watching Motivates Change: Uses and Gratifications of Streaming Video Viewers Challenge Traditional TV Research." *Convergence* 26, no. 1 (2020): 82–101.

Striphas, Ted. "Algorithmic Culture." *European Journal of Cultural Studies* 18, nos. 4–5 (2015): 395–412.

Sunstein, Cass R. *Republic.com 2.0*. Princeton, NJ: Princeton University Press, 2007.

Surowiecki, James. *The Wisdom of Crowds: Why the Many Are Smarter Than the Few*. London: Abacus, 2005.

Thaler, Richard H., and Cass B. Sunstein. *Nudge: Improving Decisions about Health, Wealth, and Happiness*. New Haven, CT: Yale University Press, 2008.

Tintarev, Nava, and Judith Masthoff. "Explaining Recommendations: Design and Evaluation." In *Recommender Systems Handbook*, 2nd rev. ed., edited by Francesco Ricci, Lior Rokach, and Bracha Shapira, 353–82. New York: Springer, 2015.

Tkalcic, Marko, and Li Chen. "Personality and Recommender Systems." In *Recommender Systems Handbook*, 2nd rev. ed., edited by Francesco Ricci, Lior Rokach, and Bracha Shapira, 715–39. New York: Springer, 2015.

Tryon, Chuck. *On-Demand Culture: Digital Delivery and the Future of Movies*. New Brunswick, NJ: Rutgers University Press, 2013.

Toff, Benjamin, and Rasmus Kleis Nielsen. "'I Just Google It': Folk Theories of Distributed Discovery." *Journal of Communication* 68, no. 3 (2018): 636–57.

Towse, Ruth. *A Textbook of Cultural Economics*. Cambridge: Cambridge University Press, 2010.

Turow, Joseph. *Breaking Up America: Advertisers and the New Media World*. Chicago: University of Chicago Press, 1997.

———. *The Daily You: How the New Advertising Industry Is Defining Your Identity and Your Worth*. New Haven, CT: Yale University Press, 2011.

Urrichio, William. "Data, Culture and the Ambivalence of Algorithms." In *The Datafied Society: Studying Culture through Data*, edited by Mirko Tobias Schäfer and Karin van Es, 125–37. Amsterdam: Amsterdam University Press, 2017.

Vanderbilt, Tom. *You May Also Like: Taste in an Age of Endless Choice*. London: Simon and Schuster, 2016.

Veblen, Thorstein. *The Theory of the Leisure Class*. Boston: Houghton Mifflin, 1973 [1899].

Verboord, Marc. "The Legitimacy of Book Critics in the Age of the Internet and Omnivorousness: Expert Critics, Internet Critics and Peer Critics in Flanders and the Netherlands." *European Sociological Review* 26, no. 6 (2010): 623–37.

Vonderau, Patrick. "The Politics of Content Aggregation." *Television and New Media* 16, no. 8 (2014): 717–33.

Wasko, Janet, and Eileen R. Meehan. "Critical Crossroads or Parallel Routes? Political Economy and New Approaches to Studying Media Industries and Cultural Products." *Cinema Journal* 52, no. 3 (2013): 150–57.

Webster, James G. *The Marketplace of Attention: How Audiences Take Shape in a Digital Age*. Cambridge, MA: MIT Press, 2014.

Weinman, Joe. *Digital Disciplines: Attaining Market Leadership via the Cloud, Big Data, Social, Mobile, and the Internet of Things*. Hoboken, NJ: Wiley, 2015.

Williams, Daniel G. *Ethnicity and Cultural Authority: From Arnold to Du Bois*. Edinburgh: Edinburgh University Press, 2006.

Williams, Raymond. *Marxism and Literature*. Oxford: Oxford University Press, 1977.

Wilson, Elizabeth J., and Daniel L. Sherrell. "Source Effects in Communication and Persuasion Research." *Journal of the Academy of Marketing Science* 21, no. 2 (1993): 102–12.

Winston, Brian. *Media, Technology, and Society: A History: From the Telegraph to the Internet*. London: Routledge, 1998.

Wolff, Michael. *Television Is the New Television: The Unexpected Triumph of Old Media in the Digital Age*. New York: Portfolio/Penguin, 2015.

Wright, David. *Understanding Cultural Taste: Sensation, Skill and Sensibility*. Basingstoke, UK: Palgrave Macmillan, 2015.

Wyatt, Robert O., and David P. Badger. "How Reviews Affect Interest in and Evaluation of Films." *Journalism Quarterly* 61, no. 4 (1984): 874–78.

Yoo, Kyung-Hyan, Ulrike Gretzel, and Markus Zanker. "Source Factors in Recommender System Credibility Evaluation." In *Recommender Systems Handbook*, 2nd rev. ed., edited by Francesco Ricci, Lior Rokach, and Bracha Shapira, 689–714. New York: Springer, 2015.

INDEX

A/B testing, 71, 94, 198

Academy Awards. *See* awards

advertising: agendas and addressees of, 40, 117; credibility of, 10, 15, 25–26, 52–55; of Netflix, 67–70, 128, 130; object of scholarly inquiry, 18, 21; personalization of, 45, 104; persuasion effects of, 35–36; remediation of, 27, 87; social needs for, 79; users' responses to, 131–35, 138–39, 148, 156, 164–65, 168. *See also* promotional rhetoric

Afrostream, 41

AI (artificial intelligence), 4, 78, 116, 151, 199, 201

algorithmic culture, 8, 13, 98, 151, 172, 178, 186, 195

All4, 142, 147

Amatriain, Xavier, 74, 86–90, 105

Amazon: business model of, 11, 103, 112–13; competition with Netflix, 68, 203; content of, 41–42; design and usability of, 55; employment practices of, 200; marketing myths of, 117–18, 198; market leader, 7, 30–33, 62, 104, 182, 188–89; new intermediary functions of, 190; personalization strategies of, 38–40, 45, 48–52; recommender system of, 71; users' responses to, 124, 126–27, 136, 139, 145, 147

AMC, 108

Apple, 104, 113, 198, 218n11

Apple TV+, 113

Arnold, Matthew, 4, 8, 38, 58–60, 99, 193

Arrested Development, 154

Arrow, 169

Atlantic, 3, 80, 110

automation, 22, 98, 121, 199–203, 206

AVOD (ad-supported video on demand), 6, 12–13, 220n26

awards, 46, 56, 63, 75, 82, 87, 101, 175; Academy Awards, 54, 65, 75, 82; Emmy Awards, 36, 65

awareness effects, 70, 74, 101, 131, 135–40, 148, 185, 241n53, 246n7

Basilico, Justin, 74, 86–90, 105

Bayesian networks, 12, 43, 117

BBC, 106, 108, 174

BBC iPlayer, 124–26, 142, 147, 212

Because You Watched function/row, 52, 63, 80, 83–84, 146, 194

Benkler, Yochai, 181

Bezos, Jeff, 11, 220n24

BFI (British Film Institute), 172

BFI Player, 6–7, 11–12, 41–42, 51, 60–61, 125, 189, 195, 212

bias: cognitive 35, 46, 49, 88, 96, 99; of critics or scholars, 176, 197; cultural, 110, 116, 134, 200; scholarly objections to, 4, 43; taste, 92–93; unconscious, 57, 253n53

big data, 3, 5, 15, 18, 53, 66, 136, 185, 195, 203, 207; the mythology of, 9, 21, 96–121

billboards, 39, 125, 128, 130, 194, 210–11, 213

binge-watching, 2, 4, 33, 47, 63, 75, 105, 130, 171, 190, 245n1

BitTorrent, 190

black box, 3, 110, 117, 119

Founded in 1893,
UNIVERSITY OF CALIFORNIA PRESS
publishes bold, progressive books and journals
on topics in the arts, humanities, social sciences,
and natural sciences—with a focus on social
justice issues—that inspire thought and action
among readers worldwide.

The UC PRESS FOUNDATION
raises funds to uphold the press's vital role
as an independent, nonprofit publisher, and
receives philanthropic support from a wide
range of individuals and institutions—and from
committed readers like you. To learn more, visit
ucpress.edu/supportus.